SpringerBriefs in Statistics

W9-DHR-491

FOR REVIEW

S. EJAZ AHMED
BOOK REVIEW EDITOR
Technometrics

More information about this series at http://www.springer.com/series/8921

Ivan Nagy · Evgenia Suzdaleva

Algorithms and Programs of Dynamic Mixture Estimation

Unified Approach to Different Types of Components

 Springer

Ivan Nagy
Department of Signal Processing
Institute of Information Theory and
  Automation of the Czech Academy of
  Sciences and Czech Technical University
  in Prague
Prague
Czech Republic

Evgenia Suzdaleva
Department of Signal Processing
Institute of Information Theory and
  Automation of the Czech Academy of
  Sciences
Prague
Czech Republic

ISSN 2191-544X   ISSN 2191-5458   (electronic)
SpringerBriefs in Statistics
ISBN 978-3-319-64670-1   ISBN 978-3-319-64671-8   (eBook)
DOI 10.1007/978-3-319-64671-8

Library of Congress Control Number: 2017947851

© The Author(s) 2017
This work is subject to copyright. All rights are reserved by the Publisher, whether the whole or part
of the material is concerned, specifically the rights of translation, reprinting, reuse of illustrations,
recitation, broadcasting, reproduction on microfilms or in any other physical way, and transmission
or information storage and retrieval, electronic adaptation, computer software, or by similar or dissimilar
methodology now known or hereafter developed.
The use of general descriptive names, registered names, trademarks, service marks, etc. in this
publication does not imply, even in the absence of a specific statement, that such names are exempt from
the relevant protective laws and regulations and therefore free for general use.
The publisher, the authors and the editors are safe to assume that the advice and information in this
book are believed to be true and accurate at the date of publication. Neither the publisher nor the
authors or the editors give a warranty, express or implied, with respect to the material contained herein or
for any errors or omissions that may have been made. The publisher remains neutral with regard to
jurisdictional claims in published maps and institutional affiliations.

Printed on acid-free paper

This Springer imprint is published by Springer Nature
The registered company is Springer International Publishing AG
The registered company address is: Gewerbestrasse 11, 6330 Cham, Switzerland

Acknowledgements

This research was supported by the project GAČR GA15-03564S.

Contents

Chapter 1
Introduction

1.1 On Dynamic Mixtures

Mixture models are known to have the ability to describe a rather wide class of real systems. They have the property of universal approximation which theoretically means that with a sufficiently large number of components, they are able to model an arbitrary system showing signs of a multimodal, nonlinear behavior, see, e.g., [1]. The area of their application is really great (industry, engineering, social fields, medicine, transportation, etc.) [2–7], and there is no particular need to introduce them in detail. However, it should be highlighted which properties of mixtures are the subject of interest in the presented book.

Any mixture model is composed of two parts—components and a switching model. Components describe different modes of behavior of the modeled system. Switching the components is modeled as a discrete random variable described by the categorical distribution. This variable is called the pointer, values of which represent labels of individual components (using the terminology from [8, 31] adopted in this book). At each time instant the pointer indicates the currently active component.

As it is seen from the title, the book focuses on dynamic mixtures. By dynamic mixtures we mean mainly a mixture with the dynamic switching model. It means that switching depends on the value of the last active component. The components can be either static (i.e., without delayed values of the modeled variable in the condition) or dynamic.

The dynamic switching of components may not always exist. Let us say, for example, that a subject of modeling is the severity of traffic accidents with the values: "fatal", "severe injury", "light injury" and "property damage". In this case, the severity can be used as the pointer variable, which has four possible values. Components could describe various data accompanying the accident—"speed", "weather conditions", "visibility on the road", "road slipperiness", etc. Notice that the dynamic model is not suitable here, since there is no dependence in switching the accident severity.

© The Author(s) 2017

I. Nagy and E. Suzdaleva, *Algorithms and Programs of Dynamic Mixture Estimation*, SpringerBriefs in Statistics, DOI 10.1007/978-3-319-64671-8_1

In contrast to that, some variables cannot switch rapidly but they are growing or declining. For example, if we want to model the occurrence of a certain disease in some location, the epidemy is unlikely to appear suddenly. It needs time to spread. At the same time, the probability of the disease is higher if the disease has been already detected in the neighborhood. Thus, the occurrence of the disease can be modeled dynamically as it depends on the last occurrence. If the regions affected by the disease are taken as the pointer values, the dynamic pointer model is suitable in this case. Construction of the model is not limited only by the last pointer value. It can include more delayed values if they bring significant information.

To use the dynamic mixture in practice, special forms of components should be chosen. In this book, we will consider components represented by normal regression, categorical and state-space models. Several examples give an idea of the application of each type of component.

Mixtures of normal components are probably most often used in literature, see, e.g., [18, 19]. For instance, such a mixture can be applied to modeling the driving style and the fuel consumption of a car. The fuel consumption is changing according to the driving style used. We will restrict the styles to be: "fuel efficient", "medium" and "sporty". We will use driving style as the pointer variable with three possible values. The dynamic pointer is suitable because we suppose that a driving style is changing dynamically according to the previous style. Normal components describe data measured on a driven car (e.g., fuel consumption, speed, acceleration, gas pedal position, etc.). Characteristics of normal distributions are switching in dependence on the active style of driving. The data-based approach discussed in the book uses the measured data for estimation of the driving style, which is not measurable.

Mixtures of categorical components are directed at a description of discrete data [9, 10]. They could find their application in marketing. For example, a mobile network operator evaluates customer satisfaction supposed to be "low", "middle" and "high". The evaluation is performed using an online questionnaire. The questionnaire collects discretized data of customers, for example, age ("junior", "adult", "senior"), monthly amount spent on a mobile phone tariff ("low", "middle", "high"), number of minutes called ("low", "middle", "high"), etc. Here, the pointer variable stands for customer satisfaction, which is estimated from data. The switching among components is evidently dynamic (if customers were fully satisfied, they hardly become totally unsatisfied).

Mixtures of linear state-space components are also a widespread tool of system modeling [27–29, 33]. For instance, they are found to be suitable for the description of car queue length at a crossroad. Within one state-space component, the queue length depends on the previous length and it also influences the measurable traffic intensity. Both the queue length and the intensity are switching in dependence on the quality of traffic service. The dynamic pointer can serve here to express the quality of traffic service also known as level of service (LOS). If the LOS is supposed to have six possible values ("entirely free flow", "reasonably free flow", "stable flow", "approaching unstable flow", "unstable flow", "congestion"), a mixture of six components is used.

Obviously, these are only examples illustrating the idea of a selected mixture application. The discussed issue is not limited neither to the applications nor the component types. However, a unified approach for the estimation of the component types (upon data modeled) can be an essential advantage in practice. An attempt of such an approach is proposed in the presented book.

Difference from other books

An up-to-date market provides us with a variety of books on the Bayesian estimation of mixture models. Surely the question arises whether the presented book brings something different. Let us see what kind of books are available.

The literature survey performed says that recent books on mixtures can be conditionally divided among several categories according to the most frequently used approaches. One of them is based on the expectation-maximization (EM) algorithm [16, 17], which is a well-known iterative method for the estimation of a parameterized model from a data sample. Its idea is to start working from some initial guess of parameters and values of unmeasured variables and to alternately estimate unknown variables based on fixed values of parameters and then the parameters on the base of fixed values of unknown variables. The process is iterative and thus it can take a considerable time for computations. Now, in the age of powerful computers, it should not be a problem. However, a fixed time of convergence is not always guaranteed: it strongly depends on initial conditions. Nevertheless, this famous algorithm is rather widespread and successful as to mixture estimation application fields, which is reported in such books as [11, 13–15, 23].

An alternative technique is the variational Bayes (VB) approach [24, 26]. This excellent and very universal algorithm indirectly approaches the likelihood function maximization and provides estimates of both model parameters and values of unknown variables. The method constructs a lower bound of the log-likelihood and iteratively improves its solution. The iterations are done again by means of the EM algorithm. The number of available books on the subject which have been published is not too high [25]. However, an extensive online VB repository can be found at www.variational-bayes.org by M. Beal and Z. Ghahramani.

Another significant group of publications is presented by Markov chain Monte Carlo (MCMC) methods used for mixtures. Among recent books in this field, the most famous is probably [12], and the topic is also touched lightly in [13, 14].

A lot of papers, reports, and online tutorials based on all of these three groups of methods can be found and there is no necessity to enumerate them here. But it is important to mention again a difference in terminology. The common terminology in the above approaches uses the *markov switching model* instead of the *dynamic pointer*, as it is stated in this book. This leads us to referring to one more group of algorithms, which has introduced the term of the pointer and which is supported in the presented book.

This group represents the recursive Bayesian estimation algorithms primarily based on papers [30] and [8]. The first of them deals with the estimation of normal regression models and the second one with a mixture of them and a static pointer without dynamic switching the components (see the example on accident severity in

Sect. 1.1). A systematic collection of algorithms of this formalism is presented in the book [31], which is the most significant publication in this specific area. It deals with a static pointer model too but what is important in the light of the approach discussed in the presented work, book [31] also considers recursive algorithms for categorical distributions. The dynamic pointer model for the mixture of normal components is introduced in [32] and later for the state-space components in [33].

Publications of the last specific group create the basis on which this book is built. The approach used in them is not the main stream formalism. However, it possesses a series of advantages in terms of derivations and computations. First, the approach is unified for components described by distributions with conjugate prior probability density functions. It requires several heuristic assumptions on distributions that can bring limitations in some cases (e.g., for nonnegative data, restricted data, etc.). However, in appropriate applications the assumptions are more than valid. Then, computations are reduced to algebraic updates of distribution statistics with a single easily computable approximation. It means that they are free of iterative processes and present a one-pass estimator for each type of component. A limitation of iterative techniques like an EM algorithm is always a convergence of algorithms. In contrast to that, the recursive one-pass estimators are not limited by convergence in this sense at all (but they need an appropriate initialization for assumed distributions). It explains why other approaches on the mixtures mentioned above are not relevant for the presented book. Recursive algorithms used in the book guarantee a fixed time of computations that depends only on the number of arriving measurements. For instance, predicting the fuel consumption of a driven car requires us to work with a very short period of measuring. It means that the algorithm should always compute the predicted value of consumption till the next period. This might be a good motivator in terms of practical use. The application is also simplified by the point estimators in the algorithms instead of a full Bayesian analysis.

A systematic collection of recursive algorithms of the dynamic mixture estimation for various types of components is still missing in the literature. This book fills this gap. It brings a unified scheme of constructing the estimation algorithm of mixtures with components whose statistics are reproducible (or their reproductivity is easily approximated). This is one of the main contributions of the book. The unified Bayesian approach enables us to have only formal differences in algorithms (for normal and categorical components in the current edition) due to statistics and forms of updates. However, the structure of the algorithms stays constant. In the case of state-space components it changes a bit. However, not significantly: the main features of the estimation keep the supported unified form.

The presented algorithms are enriched by codes of programs. Codes are implemented in Scilab 5.5.2 (www.scilab.org), which is known to be a powerful free and open source programming environment for engineering computations. This is another strong contribution of the book, which brings the following benefits.

Open source codes

Codes are primarily organized so that they have the main program, which uses further subroutines in dependence on the components' type chosen by a user. Both functions and executable procedures are provided. They are available as a text that can be copied directly into Scilab, which is free to download and also online as well. All programs are completely open and editable.

Detailed comments

Detailed comments are given to all programs directly in the code. Most programs are also additionally explained in the respective chapters of the book.

Adaptation to specific tasks

The codes presented are to be used with simulated data, both with simulation and estimation programs. This is done, first, to make programs more transparent and universal, and second, to give the user the possibility to check the correctness of codes. Naturally the codes can be tailored to other specific tasks according to the users' option. Two case studies are described as well. They demonstrate the efficiency of estimation.

Learning with the code

Engineers and researchers often use Scilab for solving specific tasks. Working with an open source code (i.e., editing it and correcting) gives an individual the unique possibility to become a better programmer. University teachers can also use open source codes as examples for students either for exercises or for advanced work. It allows students to easily study the basics of programming in Scilab and make their own software.

Free download

Using codes in Scilab can be suitable for people who do not possess proprietary software licenses for any reason (students, researchers on a home network, etc.).

However, it should be noted that the book does not aim at performing the analysis of execution efficiency of algorithms. The given codes should serve for illustrating the theory and demonstrating the functionality of the included algorithms. In this way, the target group of readers is not formed by professional programmers but rather by PhD students and researchers interested in developing theoretical algorithms in recursive mixture estimation and in practical application of the algorithms.

1.2 General Conventions

Selected basic notations used in the text are provided below in alphabetical order. Notations that require a more detailed explanation are introduced directly in the respective chapters.

- All modeled variables are generally column vectors.
- The conditional pdf for categorical random variables C, D with realizations c, d is denoted

$$f_{C|D}(c|d) \equiv f(C = c|D = d).$$

- The conditional pdf for continuous random variables A, B with realizations a, b is

$$f_{A|B}(a|b) \equiv f(a|b).$$

- $\equiv$—equality by definition
- $\propto$—equality up to the normalization constant
- The mixed distribution of continuous variables A, B with realizations a, b and categorical variables C, D with realizations c, d is denoted by

$$f_{A,C|B,D}(a, c|b, d) \equiv f(a, C = c|b, D = d).$$

- u_t—control input
- t—discrete time instant
- m_c—number of mixture components
- y_t—output random variable measured at the discrete time instant t
- c_t—pointer
- pdf—probability density function or probability function
- $f(\cdot|\cdot)$—probability density function or probability function
- x_t—unobservable state to be estimated

Layout

The layout of the book is organized in the following way. Chapter 2 is a preparative part of the book. It provides the specific theoretical background by introducing individual models (normal regression, categorical, and state-space) used in chapters on the mixture estimation. This specific theory is necessary for understanding the text further. Nevertheless, it should not be forgotten that a reader is supposed to be familiar with basics of statistical distributions and the Bayesian theory. Chapter 2 also recalls the existing algorithms of Bayesian estimation of the mentioned individual models.

Chapter 3 is devoted to the unified formulation of the dynamic mixture estimation problem, general for all included types of components. The chapter introduces general forms of components and a model of their switching, indicates main subtasks of the considered mixture estimation, and finally outlines the key steps of the recursive algorithm. The last remains unified for all considered types of components. Section 3.3 of this chapter deals with the problem of mixture prediction, substantial in questions of validation.

Details of the estimation algorithm are discussed in Chap. 4, which specifies the unified Bayesian approach for the dynamic mixture of normal regression models from Sect. 2.1, categorical components from Sect. 2.2, and the state-space ones from Sect. 2.3. An open source program with a simple example for each type of component is given.

Chapter 5 presents open source codes of programs with the implementation of discussed algorithms. The chapter provides the main program, which should be set in dependence on the considered task and specific subroutines. All of the codes are given with a detailed description and comments inside programs.

Chapter 6 demonstrates important features and the estimation possibilities of the discussed algorithms with the help of numerous experiments.

Two chapters in the Appendix are available for a reader. The first (Appendix A in Chap. 7) provides different specific knowledge useful for derivations of the important relations used in the book and also gives some derivations. Appendix B in Chap. 8 represents the open codes for supporting the auxiliary programs necessary for using the algorithms from Chap. 5 including the data simulation programs. All of the programs are available online.

Chapter 2
Basic Models

The dynamic mixture estimation discussed in the book requires a reader to be familiar with the basic single models used as mixture components. The presented edition operates with the regression model, the categorical model, and the state-space model. This chapter recalls the Bayesian estimation algorithms existing for these models. This specific theoretical background helps a reader be effective and fast oriented in the subsequent text. However, a knowledge of the basics of statistical distributions and the Bayesian theory is assumed.

2.1 Regression Model

A well-known regression model is one of the most frequently used system descriptions. It is used for modeling a continuous output variable which is supposed to be linearly dependent on delayed output values and other present or also delayed variables. They can include optional inputs which control the output and external variables which influence the output but cannot be affected. An important element of the model can also be an absolute term of the model (constant) which represents a nonzero mean value of the modeled output.

The regression model has two parts: (i) deterministic, which is a difference equation on measured data and (ii) stochastic, which is represented by a noise term. Under the assumption of stationarity, it is a stochastic sequence whose elements are i.i.d. (independent and identically distributed) with zero expectations and constant variances (covariance matrices in a multivariate case).

The most frequently used distribution for noise is the normal one. Belonging to the exponential family, it can be easily estimated even in its multivariate form, using the conjugate Gauss-inverse-Wishart distribution (GiW) [30, 31, 34, 35]. However, other distributions belonging to the exponential family can be used as well and their estimation can be made feasible.

© The Author(s) 2017
I. Nagy and E. Suzdaleva, *Algorithms and Programs of Dynamic
Mixture Estimation*, SpringerBriefs in Statistics, DOI 10.1007/978-3-319-64671-8_2

The normal regression model is represented by the following probability density function (denoted by the pdf in the text)

$$f(y_t|\psi_t, \Theta) \tag{2.1}$$

that can be defined through the difference equation

$$y_t = \psi_t'\theta + e_t = b_0 u_t + \sum_{i=1}^{n}(a_i y_{t-i} + b_i u_{t-i}) + k + e_t, \tag{2.2}$$

where

- $t = 1, 2, \ldots$ denotes discrete time instants;
- y_t is the output variable;
- u_t is the control input variable;
- $\psi_t' = [u_t', y_{t-1}', u_{t-1}', \ldots, y_{t-n}', u_{t-n}', 1]$ is the regression vector;
- n is the memory length;
- $\Theta \equiv \{\theta, r\}$ are parameters, where

  – $\theta = [b_0, a_1, b_1, \ldots, a_n, b_n, k]$ is a collection of regression coefficients,
  – and r is the constant variance of the normal noise e_t with the zero expectation.

If y_t is a vector, it is the multivariate case and the parameters are matrices of appropriate dimensions. An example of a two-dimensional output and a three-dimensional input is given below. The model is

$$\begin{bmatrix} y_{1;t} \\ y_{2;t} \end{bmatrix} = \begin{bmatrix} a_{11} & a_{12} \\ a_{21} & a_{22} \end{bmatrix} \begin{bmatrix} y_{1;t-1} \\ y_{2;t-1} \end{bmatrix} + \begin{bmatrix} b_{11} & b_{12} & b_{13} \\ b_{21} & b_{22} & b_{23} \end{bmatrix} \begin{bmatrix} u_{1;t} \\ u_{2;t} \\ u_{3;t} \end{bmatrix} + \begin{bmatrix} k_1 \\ k_2 \end{bmatrix} + \begin{bmatrix} e_{1;t} \\ e_{2;t} \end{bmatrix}, \tag{2.3}$$

where $[y_{1;t}, y_{2;t}]'$ are entries of the vector y_t. Corresponding matrices with regression coefficients $a_{11}, a_{12}, \ldots, b_{11}, b_{12}, \ldots$ and k_1, k_2 enter the collection θ. The noise has zero expectation $[0, 0]'$ and the constant (time invariant) covariance matrix

$$r = \begin{bmatrix} r_{11} & r_{12} \\ r_{21} & r_{22} \end{bmatrix}. \tag{2.4}$$

2.1.1 Estimation

The estimation of the model (2.1) obeys Bayes rule (7.25). According to [8, 30], the model (2.1) is rewritten as

$$f(y_t|\psi_t, \Theta) = (2\pi)^{-k_y/2}|r|^{-1/2}\exp\left\{-\frac{1}{2}tr\left(r^{-1}\begin{bmatrix} -I \\ \theta \end{bmatrix}' D_t \begin{bmatrix} -I \\ \theta \end{bmatrix}\right)\right\}, \tag{2.5}$$

where k_y denotes the dimension of the vector y_t; tr is a trace of the matrix; I is the unit matrix of the appropriate dimension and

$$D_t = \begin{bmatrix} y_t \\ \psi_t \end{bmatrix} \begin{bmatrix} y_t \\ \psi_t \end{bmatrix}' \tag{2.6}$$

is the so-called data matrix at time t. For a normal model (2.1) or in the rewritten form (2.5), the conjugate prior GiW pdf has the form

$$f(\Theta|d(t-1)) \propto |r|^{-0.5\kappa_{t-1}} \exp\left\{ -\frac{1}{2} tr \left(r^{-1} \begin{bmatrix} -I \\ \theta \end{bmatrix}' V_{t-1} \begin{bmatrix} -I \\ \theta \end{bmatrix} \right) \right\} \tag{2.7}$$

with the recomputable statistics V_{t-1}, which is the information matrix, and κ_{t-1}, which is the counter of used data samples. Notation $d(t)$ corresponds to the data collection up to the time t, i.e., $d(t) \equiv \{d_0, d_1, \ldots, d_t\}$, where $d_t = \{y_t, u_t\}$, and d_0 is the prior information. It means that $d(t-1) \equiv \{d_0, d_1, \ldots, d_{t-1}\}$. After substituting (2.5) and (2.7) in (7.25), these statistics are recursively updated starting from chosen initial statistics V_0 and k_0 as follows:

$$V_t = V_{t-1} + \begin{bmatrix} y_t \\ \psi_t \end{bmatrix} \begin{bmatrix} y_t \\ \psi_t \end{bmatrix}' = V_{t-1} + \begin{bmatrix} \underbrace{y_t y_t'}_{D_y} & \underbrace{y_t \psi_t'}_{D'_{y\psi}} \\ \underbrace{\psi_t y_t'}_{D_{y\psi}} & \underbrace{\psi_t \psi_t'}_{D_\psi} \end{bmatrix}, \quad \kappa_t = \kappa_{t-1} + 1, \tag{2.8}$$

whereupon the updated matrix V_t is partitioned similarly to (2.8) keeping the appropriate dimensions, i.e.,

$$V_t = \begin{bmatrix} V_y & V'_{y\psi} \\ V_{y\psi} & V_\psi \end{bmatrix}. \tag{2.9}$$

2.1.2 Point Estimates

The updated statistics V_t is partitioned as it is shown in (2.8) with dimensions given by y_t and ψ_t. Then the point estimates of the regression coefficients θ and of the noise covariance matrix r are computed respectively

$$\hat{\theta}_t = V_\psi^{-1} V_{y\psi} \quad \text{and} \quad \hat{r}_t = \frac{V_y - V'_{y\psi} V_\psi^{-1} V_{y\psi}}{\kappa_t}. \tag{2.10}$$

Detailed information is available in [8, 30]. See also details in Appendix 7.9.1.

2.1.3 Prediction

The one-step output prediction with the regression model (2.1) is defined through the predictive pdf $f\,(y_t|u_t,d\,(t-1))$. It can generally be computed in the following way:

$$f\,(y_t|u_t,d\,(t-1)) = \int_{\Theta^*} f\,(y_t,\Theta|u_t,d\,(t-1))\,d\Theta =$$

$$= \int_{\Theta^*} f\,(y_t|\psi_t,\Theta)\,f\,(\Theta|d\,(t-1))\,d\Theta, \qquad (2.11)$$

where the first pdf inside the integral is the parametrized model, and the second one is the prior pdf (i.e., the posterior pdf from the last step of the estimation). The result of integration in this simple case is analytical.[1] However, integration is relatively complex.

Accepting the point rather than the full estimate gives us a considerable simplification. The point estimate $\hat{\Theta}_{t-1}$ of the parameter Θ can be introduced by substituting the Dirac delta function $\delta\left(\Theta,\hat{\Theta}_{t-1}\right)$, where $\hat{\Theta}_{t-1}=\{\hat{\theta}_{t-1},\hat{r}_{t-1}\}$, instead of $f\,(\Theta|d\,(t-1))$ into the above integral (see Appendix 7.9.3). Thus the simplified formula for prediction is obtained in the form

$$f\,(y_t|y\,(t-1)) = f\left(y_t|\psi_t,\hat{\Theta}_{t-1}\right) = N\left(\psi_t\hat{\theta}_{t-1},\hat{r}_{t-1}\right), \qquad (2.12)$$

where N ('expectation', 'variance') stands for the normal pdf.

This formula is not only very simple but it also has a very clear meaning: for prediction, take the model and substitute the point estimates instead of unknown parameters.

2.2 Categorical Model

A discrete model with the categorical distribution can be used if all involved variables are discrete or discretized. For practical reasons, this model is applicable only if the number of variables is not too high and also if the number of different values of individual variables is small. The discrete model assigns probabilities to configurations of variable values. If there are too many such configurations, the model has a very high dimension and becomes unfeasible.

The meaning of the discrete model is not the "proportionality" as it is in the regression model. It is rather in a "sorting"—"if the case is so and so, the result will belong to this category". It indicates that methods related to the discrete model are close to the classification. From this point of view, a value of the model output represents the label of the class to which the regression vector belongs.

---

[1]The result is the Student distribution [30].

The general form of the discrete model is described by the following probability function (also denoted by a pdf)

$$f(y_t = i|\psi_t = j, \Theta), \quad i \in y^*, \quad j \in \psi^*, \tag{2.13}$$

which is represented by the table of transition to the value $y_t = i$ under condition that the $\psi_t = j$, and where Θ is the parameter, and y^*, ψ^* are finite sets of integers. The regression vector ψ_t can generally include variables $[u_t, y_{t-1}, u_{t-1}, \ldots, y_{t-n}, u_{t-n}]$, where at each time instant the output y_t has m_y possible values and the input u_t has m_u possible values. However, a large number of variables in the regression vector increases the dimension of the transition table extremely. The dimension of the regression vector can be reduced to a scalar by coding its configurations $[1, 1] \rightarrow 1, [2, 1] \rightarrow 2, \ldots, [m_u, m_y] \rightarrow m_\psi$, where m_ψ is the whole number of configurations. The set of possible configurations of the regression vector ψ_t is denoted by $\psi^* \in \{1, \ldots, m_\psi\}$. Although the general form of the categorical model (2.13) is similar to the regression model pdf (2.1), here the parameter Θ is the matrix containing stationary transition probabilities $\Theta_{i|j} \; \forall i \in y^*$ and $\forall j \in \psi^*$, and it holds

$$\Theta_{i|j} \geq 0, \; \sum_{i=1}^{m_y} \Theta_{i|j} = 1, \; \forall i \in y^*, \forall j \in \psi^*, \tag{2.14}$$

which means that the sum of probabilities in each row is equal to 1. This statement is used for all discrete models throughout the text.

The regression vector ψ_t can also be kept as a vector. For instance, for $\psi_t = [u_t, y_{t-1}]$ the model (2.13) is the transition table with m_y columns and $m_y \times m_u$ rows, i.e., for $i, j \in y^*, k \in u^*$

$$f(y_t = i|u_t = k, y_{t-1} = j, \Theta) \equiv \tag{2.15}$$

	$y_t = 1$	$y_t = 2$	$\cdots$	$y_t = m_y$
$u_t = 1, y_{t-1} = 1$	$\Theta_{1|11}$	$\Theta_{2|11}$	$\cdots$	$\Theta_{m_y|11}$
$u_t = 2, y_{t-1} = 1$	$\Theta_{1|21}$	$\cdots$	$\cdots$	$\Theta_{m_y|21}$
$\cdots$	$\cdots$	$\cdots$	$\cdots$	$\cdots$
$u_t = m_u, y_{t-1} = 1$	$\Theta_{1|m_u 1}$	$\cdots$	$\cdots$	$\Theta_{m_y|m_u 1}$
$u_t = 1, y_{t-1} = 2$	$\Theta_{1|12}$	$\Theta_{2|12}$	$\cdots$	$\Theta_{m_y|12}$
$\cdots$	$\cdots$	$\cdots$	$\cdots$	$\cdots$
$u_t = m_u, y_{t-1} = 2$	$\Theta_{1|m_u 2}$	$\cdots$	$\cdots$	$\Theta_{m_y|m_u 2}$
$\cdots$	$\cdots$	$\cdots$	$\cdots$	$\cdots$
$u_t = 1, y_{t-1} = m_y$	$\Theta_{1|1m_y}$	$\Theta_{2|1m_y}$	$\cdots$	$\Theta_{m_y|1m_y}$
$\cdots$	$\cdots$	$\cdots$	$\cdots$	$\cdots$
$u_t = m_u, y_{t-1} = m_y$	$\Theta_{1|m_u m_y}$	$\cdots$	$\cdots$	$\Theta_{m_y|m_u m_y}$

where each $\Theta_{i|kj}$ is the probability of the value $y_t = i$ conditioned by $u_t = k$ and $y_{t-1} = j$ and the statement (2.14) holds.

2.2.1 Estimation

According to [31], estimation of the discrete model (2.13) with the help of the Bayes rule (7.25) is based on using the conjugate prior Dirichlet distribution with the recursively updated statistics. The model (2.13) is written in the so-called product form [31]

$$f\left(y_t = i | \psi_t = j, \Theta\right) = \prod_{i \in y^*, j \in \psi^*} \left(\Theta_{i|j}\right)^{\delta(i,j;\, y_t, \psi_t)}, \quad i \in y^*, \quad j \in \psi^*, \qquad (2.16)$$

where the Kronecker delta function $\delta(i, j;\, y_t, \psi_t) = 1$, when $i = y_t$ and $j = \psi_t$ and it is equal to 0 otherwise. The conjugate prior Dirichlet pdf for the model (2.13) has the form

$$f(\Theta | d(t-1)) \propto \prod_{i \in y^*, j \in \psi^*} \left(\Theta_{i|j}\right)^{(\nu_{i|j})_{t-1}-1}, \quad i \in y^*, \quad j \in \psi^*, \qquad (2.17)$$

where the statistics ν_{t-1} is a matrix of the same dimension as (2.13) and $(\nu_{i|j})_{t-1}$ are its entries.

Substituting the model (2.16) and the prior pdf (2.17) into (7.25) gives updating the entries of the statistics ν_{t-1}

$$(\nu_{i|j})_t = (\nu_{i|j})_{t-1} + \delta(i, j;\, y_t, \psi_t), \quad \forall i \in y^*, \quad \forall j \in \psi^*, \qquad (2.18)$$

with some chosen initial statistics ν_0. In practice, it means that the statistics counts occurrences of the combinations of values of y_t and ψ_t.

2.2.2 Point Estimates

The point estimate of the parameter Θ is obtained by normalizing the statistics ν_t

$$(\hat{\Theta}_{i|j})_t = \frac{(\nu_{i|j})_t}{\sum_{k=1}^{m_y} (\nu_{k|j})_t}, \quad i \in y^*, \quad j \in \psi^*. \qquad (2.19)$$

The detailed information is available in [31]. See also Appendix 7.8.

2.2.3 Prediction

For the output prediction with the discrete model (2.13) it holds

$$f(y_t = i | u_t = k, d(t-1)) = \int_0^1 \underbrace{f(y_t = i | \psi_t = j, \Theta)}_{(2.13)} \underbrace{f(\Theta | d(t-1)) \, d\Theta}_{(2.17)} = (\hat{\Theta}_{i|j})_{t-1},$$
(2.20)

which is the expectation of the Dirichlet distribution and the regression vector ψ_t is comprised from values of the input u_t and past data $d(t-1)$.

2.3 State-Space Model

A state is a variable whose statistical properties are fully determined by its last value and by the actual control input (if present). Its prediction depends only on its last value, not on the whole history of its evolution. Very often the state variable cannot be measured, i.e., it is modeled and estimated. For its description, the state-space model is suitable. It consists of two parts: the *state model* describes the state evolution in dependence on its last value and the control. The *output model* (sometimes called *the measurement model*) reflects the effect of the state and possibly of the control on the output variable.

The general form of the state-space model is presented in the following two pdfs:

$$f(x_t | x_{t-1}, u_t) \quad \text{state model,} \tag{2.21}$$

$$f(y_t | x_t, u_t) \quad \text{output model,} \tag{2.22}$$

where x_t denotes the unobservable state to be estimated. The linear normal state-space model described by these pdfs can be written with the help of the equations

$$x_t = M x_{t-1} + N u_t + F + \omega_t, \tag{2.23}$$

$$y_t = A x_t + B u_t + G + v_t, \tag{2.24}$$

where M, N, F, A, B and G are matrices of parameters of appropriate dimensions supposed to be known; ω_t and v_t are the process as well as the measurement Gaussian white noises with zero expectations and covariance matrices R_ω and R_v respectively, which are usually supposed to be known.

2.3.1 State Estimation

Bayesian state estimation (that can be found in various sources, e.g., [30, 36], etc.) operates with the prior state pdf $f(x_{t-1}|d(t-1))$ and the state-space model (2.21)–(2.22) to obtain the posterior state pdf $f(x_t|d(t))$ via the recursion

$$f(x_t|d(t-1)) = \int_{x^*} f(x_t|x_{t-1}, u_t) f(x_{t-1}|d(t-1)) dx_{t-1}, \qquad (2.25)$$

$$f(x_t|d(t)) = \frac{f(y_t|x_t, u_t) f(x_t|d(t-1))}{f(y_t|u_t, d(t-1))}, \qquad (2.26)$$

with

$$f(y_t|u_t, d(t-1)) = \int_{x^*} f(y_t|x_t, u_t) f(x_t|d(t-1)) dx_t. \qquad (2.27)$$

The recursion starts with the chosen prior pdf $f(x_0|d(0))$ representing prior knowledge about the state.

Substituting the linear form (2.23)–(2.24) into the above recursion gives us the famous Kalman filter [30, 37–39]. For linear normal models and the normal initial state given by the prior pdf

$$f(x_{t-1}|d(t-1)) = N(\hat{x}_{t-1|t-1}, R_{t-1|t-1}), \qquad (2.28)$$

the state description during its evolution stays normal. With this prior pdf and the models (2.23) and (2.24), the Kalman filter can be presented as follows.

Prediction

$$\text{state prediction } \hat{x}_{t|t-1} = M\hat{x}_{t-1|t-1} + Nu_t + F, \qquad (2.29)$$
$$\text{prediction of state covariance } R_{t|t-1} = R_\omega + MR_{t-1|t-1}M', \qquad (2.30)$$

Filtration

$$\text{output prediction } \hat{y}_t = A\hat{x}_{t|t-1} + Bu_t + G, \qquad (2.31)$$
$$\text{noise covariance update } R_y = R_v + AR_{t|t-1}A', \qquad (2.32)$$
$$\text{update of the state covariance } R_{t|t} = R_{t|t-1} - R_{t|t-1}A'R_y^{-1}AR_{t|t-1}, \qquad (2.33)$$
$$\text{Kalman gain } K_g = R_{t|t}A'R_v^{-1}, \qquad (2.34)$$
$$\text{state correction } \hat{x}_{t|t} = \hat{x}_{t|t-1} + K_g(y_t - \hat{y}_t), \qquad (2.35)$$

where $\hat{y}_t$ denotes the point prediction of the output obtained by substituting the current point estimates of the state. The point estimates of the state are given by the values of $\hat{x}_{t|t}$ after the state correction.

Remark In this way, this chapter summarizes the available knowledge on basic models which are used further for modeling the components and the pointer. Now, it is necessary to introduce the dynamic mixture model which is explained in the next chapter.

Chapter 3
Statistical Analysis of Dynamic Mixtures

3.1 Dynamic Mixture

A mixture model considered in this book consists of a set of m_c components, which can be either regression models (2.1), categorical models (2.13), or state-space models (2.21)–(2.22). In general form the component model is denoted by the pdf

$$f(y_t|\psi_t, \Theta, c_t = i), \tag{3.1}$$

where $i \in \{1, 2, \ldots, m_c\}$, which will be used as the set c^* for brevity and c_t is the discrete variable called the pointer [31]. A value of the pointer c_t denotes the component active at the current time instant t. The dynamic pointer is described by the categorical model of the form (2.13) with notations reserved only for the pointer description. It is

$$f(c_t = i|c_{t-1} = j, \alpha) \equiv \tag{3.2}$$

	$c_t = 1$	$c_t = 2$	$\cdots$	$c_t = m_c$
$c_{t-1} = 1$	$\alpha_{1|1}$	$\alpha_{2|1}$	$\cdots$	$\alpha_{m_c|1}$
$c_{t-1} = 2$	$\alpha_{1|2}$		$\cdots$	
$\cdots$	$\cdots$	$\cdots$	$\cdots$	$\cdots$
$c_{t-1} = m_c$	$\alpha_{1|m_c}$		$\cdots$	$\alpha_{m_c|m_c}$

where $i, j \in c^*$, the parameter α is the $(m_c \times m_c)$-dimensional matrix, whose entries $\alpha_{i|j}$ are the probabilities of transition from $c_{t-1} = j$ to $c_t = i$ and the statement of the form (2.14) holds. Thus, it can be seen, that the discussed dynamics of the mixture expresses modeling the transition among components depending on the parameter α and the last active component, i.e., at time $t - 1$. However, in modeling a real system, the pointer can be often an unmeasurable process. This indicates a necessity to estimate the pointer value c_{t-1} too. To summarize the problem, the variables,

© The Author(s) 2017   19
I. Nagy and E. Suzdaleva, *Algorithms and Programs of Dynamic Mixture Estimation*, SpringerBriefs in Statistics, DOI 10.1007/978-3-319-64671-8_3

which are unavailable and have to be estimated at time t are

$$c_t, \ c_{t-1}, \ \Theta, \ \alpha.$$

This is a unified problem formulation general for all included types of components. Naturally, details of the estimation algorithm will be distinguished depending on the considered components (they will be specified later). However, the general approach of the Bayesian recursive estimation of dynamic mixtures will keep the unified form. It is presented below.

3.2   Unified Approach to Mixture Estimation

The mixture estimation algorithm has been developed with the help of the construction of the joint pdf of all variables that should be estimated based on measured data. The next step is the application of the Bayes rule (7.25) and the chain rule (7.26). Generally, the joint pdf of the above variables has the form

$$f(\Theta, c_t = i, c_{t-1} = j, \alpha | d(t)) \tag{3.3}$$

with $i, j \in c^*$. With the help of (7.25) and (7.26), it can be expressed as proportional to a product of the two following pdfs:

$$f(d_t, \Theta, c_t = i, c_{t-1} = j, \alpha | d(t-1))$$
$$= \underbrace{f(d_t, \Theta | c_t = i, d(t-1))}_{\text{the component part}} \underbrace{f(c_t = i, c_{t-1} = j, \alpha | d(t-1))}_{\text{the pointer part}}, \tag{3.4}$$

where the component part relates to the description of the i-th component and the pointer part—to modeling the pointer and where parameters Θ and α are assumed to be mutually independent.

3.2.1   The Component Part

The component part from (3.4) is decomposed according to the chain rule (7.26) as follows:

$$f(d_t, \Theta | c_t = i, d(t-1)) = \underbrace{f(y_t | \psi_t, \Theta, c_t = i)}_{\text{component (3.1)}} \underbrace{f(\Theta | d(t-1))}_{\text{the prior pdf}}. \tag{3.5}$$

Its utilization can be twofold, i.e.,

1. Application of the Bayes rule and the decomposition produce the posterior pdf

$$f\left(\Theta | c_t = i, d\left(t\right)\right) \propto f\left(d_t, \Theta | c_t = i, d\left(t-1\right)\right), \tag{3.6}$$

which says: take the component model, multiply it by the prior pdf (posterior from the previous step) and obtain the actual posterior pdf. This represents the task of estimation of the i-th component.

2. The integration over Θ gives

$$f\left(d_t | c_t = i, d\left(t-1\right)\right) = \int_{\Theta^*} f\left(d_t, \Theta | c_t = i, d\left(t-1\right)\right) d\Theta, \tag{3.7}$$

which is the integrated likelihood or the so-called v-likelihood [31]. It expresses the suitability of the measured data item d_t to individual components with $i \in c^*$.

3.2.2 The Pointer Part

The pointer part from (3.4) is decomposed formally in the same way as the component one, i.e.,

$$f\left(c_t = i, c_{t-1} = j, \alpha | d\left(t-1\right)\right) \propto \underbrace{f\left(c_t = i | c_{t-1} = j, \alpha\right)}_{\text{the pointer model (3.2)}} \underbrace{f\left(\alpha | d\left(t-1\right)\right)}_{\text{the prior pdf}}, \tag{3.8}$$

with $i, j \in c^*$ again with a similar twofold meaning

1. It can be used as a formula for the estimation of the parameter α under the assumption that the pointer values are measured as data (which would be performed according to [31] and (2.16)–(2.18) in this case), i.e.,

$$f\left(\alpha | d(t), c_t = i, c_{t-1} = j\right) \propto f\left(c_t = i, c_{t-1} = j, \alpha | d\left(t-1\right)\right). \tag{3.9}$$

2. The integration over α provides a computation of the v-likelihood [31], i.e.,

$$f\left(c_t = i, c_{t-1} = j | d\left(t-1\right)\right) = \int_{\alpha^*} f\left(c_t = i, c_{t-1} = j, \alpha | d\left(t-1\right)\right) d\alpha. \tag{3.10}$$

These two formulas create a starting assumption for the mixture estimation with unknown (unavailable) pointer values c_t and c_{t-1}. For the practical use of (3.9) and (3.10), these values have to be replaced by their estimates, whereupon their usage is straightforward.

3.2.3   Main Subtasks of Mixture Estimation

The above scheme is a basis for all of the following subtasks within the recursive estimation of dynamic mixtures. The subtasks include, for $i, j \in c^*$:

Pointer estimation (classification)

$$f\left(c_t = i | d\left(t\right)\right) \propto \sum_{j=1}^{m_c} \int_{\Theta^*} \int_{\alpha^*} f\left(\Theta, c_t = i, c_{t-1} = j, \alpha | d\left(t\right)\right) d\alpha d\Theta \qquad (3.11)$$

Parameter estimation

$$f\left(\Theta, \alpha | d\left(t\right)\right) \propto \sum_{i=1}^{m_c} \sum_{j=1}^{m_c} f\left(\Theta, c_t = i, c_{t-1} = j, \alpha | d\left(t\right)\right) \qquad (3.12)$$

Data prediction

$$f\left(d_t | d\left(t-1\right)\right) = \sum_{i=1}^{m_c} \sum_{j=1}^{m_c} \int_{\Theta^*} \int_{\alpha^*} f\left(d_t, \Theta, c_t = i, c_{t-1} = j, \alpha | d\left(t-1\right)\right) d\alpha d\Theta.$$
$$(3.13)$$

However, the recursive reconstruction of the joint pdf (3.3), which is one of the main goals in this book, is not feasible. The reason is the consideration of pointer values as measured data. By admitting that the pointer is unknown, we are able to marginalize. Then the model, which enters the Bayes formula gets a summation form. It can be seen from the following expression:

$$\underbrace{f\left(\Theta, \alpha | d\left(t\right)\right)}_{(3.12)} \propto \sum_{i=1}^{m_c} \sum_{j=1}^{m_c} \underbrace{\overbrace{f\left(\Theta, c_t = i, c_{t-1} = j, \alpha | d\left(t\right)\right)}^{(3.3)}}_{(3.12)} \propto$$

$$\propto \sum_{i=1}^{m_c} \sum_{j=1}^{m_c} \underbrace{f\left(d_t, \Theta, c_t = i, c_{t-1} = j, \alpha | d\left(t-1\right)\right)}_{(3.4)} =$$

$$= \underbrace{\sum_{i=1}^{m_c} \sum_{j=1}^{m_c} f\left(d_t | \Theta, c_t = i, d\left(t-1\right)\right) f\left(c_t = i, c_{t-1} = j | \alpha, d\left(t-1\right)\right)}_{\text{model}} \times \underbrace{f\left(\Theta, \alpha | d\left(t-1\right)\right)}_{\text{prior pdf}}.$$
$$(3.14)$$

In this form, the recursive use of the Bayes rule produces products of sums. The complexity of the description of the posterior pdf grows and the estimation gets unfeasible. This is the reason to simplify computations by a proper approximation. The approximation chosen is as follows [8, 40]:

1. The value $i \in c^*$ of the pointer c_t is assumed to be available, i.e., it is known, which component is currently active.
2. For the known pointer, its pdf in the form of the Kronecker delta function is introduced as follows

$$f(c_t = i|d(t)) = \delta(c_t, i), \tag{3.15}$$

   where $i \in c^*$ denotes the component assumed to be active at time t and $\delta(c_t, i) = 1$, if $c_t = i$ and 0 otherwise.
3. Since in reality the pointer value is not known, its whole pdf is estimated by the expectation [8, 40]

$$\delta(c_t, i) \rightarrow E[\delta(c_t, i)|d(t)], \tag{3.16}$$

   which gives us the estimate of the pointer distribution based on the data $d(t)$.
4. The dynamic mixture estimation requires to have both c_t and c_{t-1} also estimated. The following rearrangements are applied to obtain them. Using marginalization of the joint pdf (3.3) over the parameters Θ and α, the result is

$$\underbrace{f(c_t = i, c_{t-1} = j|d(t))}_{\text{denoted by } w_{j,i;t}} \propto \int_{\Theta^*} \int_{\alpha^*} \underbrace{f(\Theta, c_t = i, c_{t-1} = j, \alpha|d(t))}_{(3.3)} d\alpha d\Theta. \tag{3.17}$$

This is the pdf joint for c_t and c_{t-1}, which is denoted by $W_{j,i;t}$ and

$$\delta(c_t, c_{t-1}; i, j) \rightarrow E[\delta(c_t, c_{t-1}; i, j)|d(t)] = \underbrace{f(c_t = i, c_{t-1} = j|d(t))}_{W_{j,i;t}}. \tag{3.18}$$

To obtain the expectation of the pointer distribution (3.16), the pdf (3.18) is marginalized again over c_{t-1}, i.e.,

$$\underbrace{f(c_t = i|d(t))}_{\text{denoted by } w_{i;t}} \propto \sum_{j \in c^*} f(c_t = i, c_{t-1} = j|d(t)), \tag{3.19}$$

which gives the probability (weight) $w_{i;t}$ of the activity of the i-th component at time t and

$$\delta(c_t, i) \rightarrow E[\delta(c_t, i)|d(t)] = w_{i;t}, \tag{3.20}$$

and $[w_{1;t}, w_{2;t}, \ldots, w_{m_c;t}]'$ creates the weighting vector, which is the estimated pointer distribution. It becomes the prior pointer pdf at the next step of the recursion.

With this approximation scheme the estimation is feasible. At each time instant during the estimation, not only the active component (as it would be in the case of known component activities) but also all of the components are updated, each with its own probability of being active. Another helpful type of approximation is also used for a successful estimation. It consists of substituting the point estimates of parameters and the measured data item into the corresponding model, which evaluates a proximity of the current data item from individual components (see Appendix 7.9.3).

The above arguments lead to the following structure of the mixture estimation algorithm.

3.2.4 General Algorithm

1. Measuring the new data item.
2. This step should concern the evaluation of the v-likelihood of each component with the newly measured data item. The v-likelihood expresses how close the actual data items are to individual components. However, in this book we use the approximation by evaluation of the proximity of the current data item to individual components (see Appendix 7.9.3).
3. Again, the evaluation of the v-likelihood of the pointer model is similarly replaced by computing the proximity, which reflects frequencies with which individual components were active in the past.
4. Constructing the component weights as products of the mentioned proximities and the previous weights.
5. Updating statistics of all components, each with its corresponding weight.
6. Updating the statistics of the pointer model using the component weights.
7. Computing the point estimates of parameters of all models. They are used in the practical evaluation of proximities in the next step.

In this way, the general steps of the recursive mixture estimation are outlined. They keep the unified form for all types of components from Chap. 2. Specific details of the algorithm for each component type can be found in Chap. 4.

3.3 Mixture Prediction

A task of prediction based on a mixture model is not directly a part of the mixture estimation. Here it is added with the aim of the demonstration of results of the estimation. However, the prediction task is not trivial. This section provides a summary of the problem. It will concern autoregressive models only (without control and external variables). The extension to a controlled model is straightforward. However, one must not forget about the natural conditions of control (NCC) [30], see also Appendix 7.7.

Similarly, as in the task of the estimation, it is assumed that at the time instant t the available data collection consists of $d(t-1) = \{d_0, d_1, \cdots, d_{t-1}\}$ and also of the actual data item d_t, which is presented here only by y_t. Data from the future time instants $d_{t+1}, d_{t+2}, \cdots$ are unknown. The task is to estimate the output for the time instant $t + n_p$, i.e., to predict the output n_p steps ahead.

The task of the mixture prediction is twofold

1. **Pointer prediction**
   that estimates which component will be active n_p steps ahead
   This task will be considered in its general n_p steps ahead form. The pointer prediction for time t (i.e., for $n_p = 0$, equivalent to the estimation of the component which is currently active) coincides with the task of classification (assigning the actual data item d_t to one of the components). The n_p steps pointer prediction is a new feature which can be called the predicted classification. It estimates how the future data item d_{t+n_p} will be classified. The formulation of this task is as follows: determine the pdf

$$f\left(c_{t+n_p} = i | d(t)\right), \ \forall i \in c^* \tag{3.21}$$

   using only the data $d(t)$ in the condition.

2. **Data prediction**
   that generally estimates the future output of the mixture model
   This task is rather problematic due to the fact, that the components are switching during a period of prediction. This is why the task is formulated here as only the zero steps prediction, i.e., for $n_p = 0$ and is also considered for static components. This means the prediction of the data item d_t, when it has already been measured but not yet used for the mixture estimation. Such a prediction is useful for computing the so-called prediction error to evaluate the quality of the estimation. However, the use of d_t for its own prediction should be avoided, even indirectly. It means neither the updated parameter estimates nor the weighting vector w_t must be used for predicting d_t. The formulation of the task is as follows: construct the pdf

$$f(d_t | d(t-1)) \tag{3.22}$$

   using only the variables in the condition.

3.3.1 Pointer Prediction

To be able to decide how to construct the predictive pdf (3.21), it is necessary, similarly as in previous sections, to consider the joint pdf of all unknown variables. For better transparency, the two steps ahead prediction is discussed. However, the generalization is straightforward. The joint pdf for brevity denoted by $\mathcal{J}^P$ is

$$\mathcal{J}^P \;=\; f\left(c_{t+2}=k, c_{t+1}=l, c_t=i, c_{t-1}=j, \Theta, \alpha | d\,(t)\right)$$

$$\underbrace{\propto}_{\text{Bayes rule}} \; f\left(d_t, c_{t+2}=k, c_{t+1}=l, c_t=i, c_{t-1}=j, \Theta, \alpha | d\,(t-1)\right), \qquad (3.23)$$

where $k, l, i, j \in c^*$. From this pdf the predictive one (3.21) can be computed by integration over parameters Θ and α and summation over values of c_{t+1}, c_t and c_{t-1}.

This joint pdf can be factorized using the chain rule (see Appendix 7.6) under the assumption of a mutual parameter independence (see Sect. 3.2). It is obtained

$$\mathcal{J}^P = \underbrace{f\,(y_t|\psi_t, \Theta, c_t=i)}_{(3.1)}\, \underbrace{f\,(c_{t+2}=k|c_{t+1}=l, \alpha)}_{(3.2)}\, \underbrace{f\,(c_{t+1}=l|c_t=i, \alpha)}_{(3.2)}\, \underbrace{f\,(c_t=i|c_{t-1}=j, \alpha)}_{(3.2)}$$

$$\times\; \underbrace{f\,(c_{t-1}=j|d\,(t-1))}_{w_{j;t-1}}\, \underbrace{f\,(\Theta|d\,(t-1))}_{\Theta\text{prior pdf}}\, \underbrace{f\,(\alpha|d\,(t-1))}_{\alpha\text{prior pdf}}, \qquad (3.24)$$

which includes the pdfs of the components (3.1), the pointer (3.2) and the corresponding prior pdfs. Integrating (3.24) over Θ and α gives

$$\int_{\Theta^*}\int_{\alpha^*} \mathcal{J}^P\, d\alpha d\Theta = \int_{\Theta^*} f\,(y_t|\psi_t, \Theta, c_t=i)\, f\,(\Theta|d\,(t-1))\, d\Theta$$

$$\times \int_{\alpha^*} f\,(c_{t+2}=k|c_{t+1}=l, \alpha)\, f\,(c_{t+1}=l|c_t=i, \alpha)\, f\,(c_t=i|c_{t-1}=j, \alpha)\, f\,(\alpha|d\,(t-1))\, d\alpha \qquad (3.25)$$

$$\times f\,(c_{t-1}=j|d\,(t-1)),$$

which, according to Sect. 3.2.3 and using approximations by taking point estimates of parameters instead of the full pdfs, results in

$$f\left(y_t|\psi_t, \hat{\Theta}_{i;t-1}\right) \hat{\alpha}_{k|l;t-1}\hat{\alpha}_{l|i;t-1}\hat{\alpha}_{i|j;t-1}w_{j;t-1}$$

$$= \hat{\alpha}_{k|l;t-1}\hat{\alpha}_{l|i;t-1} \times f\left(y_t|\psi_t, \hat{\Theta}_{i;t-1}\right) \hat{\alpha}_{i|j;t-1}w_{j;t-1}. \qquad (3.26)$$

The summation of (3.26) over values of c_{t+1}, c_t and c_{t-1} provides

$$f\,(c_{t+2}=k|d\,(t)) \propto \sum_{l,i\in c^*} \hat{\alpha}_{k|l;t-1}\hat{\alpha}_{l|i;t-1} \sum_{j\in c^*} f\left(y_t|\psi_t, \hat{\Theta}_{i;t-1}\right) \hat{\alpha}_{i|j;t-1}w_{j;t-1}$$

$$= \sum_{l,i\in c^*} \hat{\alpha}_{k|l;t-1}\hat{\alpha}_{l|i;t-1}w_{i;t} \qquad (3.27)$$

where, according to (3.11) and Sect. 3.2.3 and Chap. 4

$$w_{i;t} = f\left(y_t|\psi_t, \hat{\Theta}_{i;t-1}\right) \sum_{j \in c^*} \hat{\alpha}_{i|j;t-1} w_{j;t-1}. \tag{3.28}$$

The verbal interpretation of the result is as follows. The two steps ahead pointer prediction from the time instant t is constructed using the actual weighting vector w_t (which is the optimal guess of the active component at time t, using the data item d_t actually measured) and predicting this vector two times with the transition table $\hat{\alpha}_{t-1}$.

Now, it is easy to extrapolate the general case of the n_p-steps pointer prediction. Again, the prediction will be given by the actual weighting vector w_t and its n_p-steps ahead prediction via the n_p-th power of the transition table $\hat{\alpha}_{t-1}$.

3.3.2 Data Prediction

Again, the derivation starts with the construction of the joint pdf from which the data prediction (3.22) can be obtained. It is

$$\mathcal{J}^D = f(d_t, c_t = i, c_{t-1} = j, \Theta, \alpha | d(t-1))$$

$$= \underbrace{f(y_t|\psi_t, \Theta, c_t = i)}_{(3.1)} \underbrace{f(c_t = i|c_{t-1} = j, \alpha)}_{(3.2)} \underbrace{f(c_{t-1} = j|d(t-1)) f(\Theta|d(t-1)) f(\alpha|d(t-1))}_{\text{prior pdfs}}$$

$$= f(y_t|\psi_t, \Theta, c_t = i) f(\Theta|d(t-1)) \times f(c_t = i|c_{t-1} = j, \alpha) f(\alpha|d(t-1)) \times w_{j;t-1}, \ \forall i, j \in c^*. \tag{3.29}$$

The integration of (3.29) over Θ and α gives

$$f(d_t|c_t = i, c_{t-1} = j, d(t-1)) = f\left(y_t|\psi_t, \hat{\Theta}_{i;t-1}\right) \times \hat{\alpha}_{i|j;t-1} w_{j;t-1} \tag{3.30}$$

and the summation over values of c_t and c_{t-1} finally provides the data predictive pdf

$$f(d_t|d(t-1)) = \sum_{i \in c^*} \hat{w}_{i;t} f\left(y_t|\psi_t, \hat{\Theta}_{i;t-1}\right), \tag{3.31}$$

where $\hat{w}_{i;t} = \sum_{j \in c^*} \hat{\alpha}_{i|j;t-1} w_{j;t-1}$ and expresses a prediction of the weight $w_{i;t}$ of the i-th component without measuring the data item y_t.

Remark In both types of prediction, the value of the weighting vector w_{t-1} is used. It is computed in the last step of the estimation according to the algorithms from Sect. 3.2.3 and Chap. 4.

Chapter 4
Dynamic Mixture Estimation

This chapter specifies the unified Bayesian approach of the recursive estimation to the dynamic mixture of normal regression models from Sect. 2.1, categorical components from Sect. 2.2 and state-space models from Sect. 2.3.

4.1 Normal Regression Components

Here a mixture of m_c normal regression models (2.2) is considered. The general pdf (3.1) from Sect. 3.1 used for these components takes the form

$$f(y_t|\psi_t, \Theta, c_t = i) = \underbrace{N(\psi_t\theta_i, r_i)}_{(2.2)}, \quad \forall i \in c^*, \tag{4.1}$$

where θ_i and r_i are the collection of regression coefficients and the noise covariance matrix (respectively) of the i-th component and $\Theta \equiv \{\theta_i, r_i\}_{i=1}^{m_c}$. Switching the activity of the components is described by the dynamic model (3.2).

Statistics update

The update of statistics of the individual normal regression model is performed according to (2.8). After the approximation outlined in Sect. 3.2.3, where the Kronecker delta function as the pointer distribution is replaced by the weight w_t, the difference of the updates is only in weighing the actual data added to the statistics by the entries of the vector w_t [8], i.e.,

$$V_{i;t} = V_{i;t-1} + w_{i;t} \begin{bmatrix} y_t \\ \psi_t \end{bmatrix} \begin{bmatrix} y_t \\ \psi_t \end{bmatrix}' \tag{4.2}$$

$$\kappa_{i;t} = \kappa_{i;t-1} + w_{i;t}, \tag{4.3}$$

© The Author(s) 2017
I. Nagy and E. Suzdaleva, *Algorithms and Programs of Dynamic Mixture Estimation*, SpringerBriefs in Statistics, DOI 10.1007/978-3-319-64671-8_4

where $V_{i;t}$ and $\kappa_{i;t}$ with the subscript i are statistics of the i-th component $\forall i \in c^*$. The statistics of the pointer model is updated similarly to the update of the individual categorical model (2.18) with the weight $W_{i,j;t}$ from (3.18) as follows:

$$v_{i|j;t} = v_{i|j;t-1} + W_{j,i;t}, \quad i, j \in c^*. \tag{4.4}$$

Point estimates

Similarly, as in the case of a single regression model, the point estimates of the parameters of each component can be evaluated according to relations (2.10). The point estimate of α is obtained by normalizing the statistics v_t according to (2.19).

Prediction

The output prediction for the i-th component is given in (2.12) based on substituting the point estimates of the parameters of the corresponding component.

To avoid sophisticated derivations of the weights $w_{i;t}$ and $W_{i,j;t}$ that can be found in other sources [32, 33] based on [8], a much more straightforward presentation of the estimation steps is given in the form of the following algorithm.

4.1.1   Algorithm

One step of the recursive estimation at time t

1. Let's have point estimates of parameters $\hat{\Theta}_{t-1} = \left\{ \hat{\Theta}_{i;t-1} \right\}_{i=1}^{m_c} = \left\{ \hat{\theta}_{i;t-1}, \right.$ $\hat{r}_{i;t-1} \right\}_{i=1}^{m_c}$ and $\hat{\alpha}_{t-1}$, which are either obtained from the previous step of the estimation according to (2.10) and (2.19) or chosen as the prior ones before the beginning of the estimation.
2. Let's have the m_c-dimensional weighting vector w_{t-1}, which is the estimated distribution of the pointer c_{t-1}, i.e., $w_{i;t-1} = f(c_{t-1} = i | d(t-1))$, obtained from the previous step of the estimation or chosen as the initial one.
3. Measure a new data item $d_t = \{y_t, u_t\}$.
4. For each component $i \in c^*$ take the current data item and the previous point of estimate of Θ and evaluate the proximity of the current data item to each component

$$m_i = f\left(y_t | \psi_t, \hat{\Theta}_{i;t-1} \right)$$

$$= (2\pi)^{-k_y/2} |\hat{r}_{i;t-1}|^{-1/2} \exp\left\{ -\frac{1}{2} \left(y_t - \psi_t \hat{\theta}_{i;t-1} \right)' \hat{r}_{i;t-1}^{-1} \left(y_t - \psi_t \hat{\theta}_{i;t-1} \right) \right\}. \tag{4.5}$$

5. Construct the weight matrix W_t, which contains the pdfs $W_{j,i;t}$ (joint for c_t and c_{t-1})

$$W_t \propto \left(w_{t-1}m'\right) . * \hat{\alpha}_{t-1} =$$

$$= \left\{ \begin{bmatrix} w_{1;t-1} \\ w_{2;t-1} \\ \cdots \\ w_{m_c;t-1} \end{bmatrix} \begin{bmatrix} m_1, m_2, \cdots, m_{m_c} \end{bmatrix} \right\} . * \begin{bmatrix} \hat{\alpha}_{1|1} & \hat{\alpha}_{2|1} & \cdots & \hat{\alpha}_{m_c|1} \\ \hat{\alpha}_{1|2} & & & \\ \cdots & \cdots & & \cdots \\ \hat{\alpha}_{1|m_c} & \hat{\alpha}_{2|m_c} & \cdots & \hat{\alpha}_{m_c|m_c} \end{bmatrix}, \quad (4.6)$$

where .∗ is a "dot product" that multiplies the matrices entry by entry.
6. Normalize the matrix W_t from (4.6) so that the overall sum of all its entries is equal to 1.
7. Perform the summation of the normalized matrix W_t over rows and obtain the weighting vector w_t (the distribution of the pointer c_t) with updated entries $w_{i;t}$ $\forall i \in c^*$ according to (3.19) as follows:

$$f\left(c_t = i | d\left(t\right)\right) = w_{i;t} \propto \sum_{j=1}^{m_c} W_{j,i;t}, \quad (4.7)$$

which gives probabilities of the activity of individual components at the current time instant t.
8. Update the statistics of each i-th component according to (4.2), (4.3) and the statistics of the pointer using (4.4).
9. Recompute the point estimates of the parameters $\hat{\Theta}_{i;t}$ of each i-th component according to (2.10).
10. Recompute the point estimate $\hat{\alpha}_t$ of the pointer model with the help of (2.19).
11. Use $\hat{\Theta}_{i;t}$, $\hat{\alpha}_t$, and w_t for the next step of the recursive estimation as the prior ones.

A simple program demonstrating the above algorithm is implemented in the open source software Scilab 5.5.2 (see www.scilab.org). It is presented below together with a detailed description and comments.

4.1.2 Simple Program

```
// Estimation of dynamic mixture with normal components
exec("Intro.sce",-1), rand('seed',139737), disp 'Running ...'

nd=1000;                        // number of data            //  1
nc=3;                           // number of components      //  2

// INITIALIZATION of simulation ---------------------------------------
thS=[-1 1 5];                   // centers of components
for i=1:nc, Sim(i).th=thS(i); end   // centers                   //  3
for i=1:nc, Sim(1).cv=.1; end   // covariances               //  4
alS=[.8 .1 .1;.1 .2 .7;.3 .1 .6];   // parameters of the pointer //  5
c(1)=1;                         // initial values           //  6
```

```
// SIMULATION
for t=2:nd
  rnd=rand(1,1,'unif');
  cus=cumsum(alS(c(t-1),:));
  c(t)=sum(rnd>cus)+1;              // pointer generation              //  7
  th=Sim(c(t)).th;
  sd=sqrt(Sim(c(t)).cv);
  y(t)=th+sd*rand(1,1,'norm');      // output generation               //  8
end

// INITIALIZATION of estimation -------------------------------------
a(1)=-3; a(2)=2; a(3)=6; r=.5;      // initial param. estimates         //  9
for i=1:nc                                                             // 10
  yy=a(i)+r*rand(1,10,'norm');      // prior data (length = 10)
  Ps=[yy; ones(1,10)]';            // regression vector
  Est(i).V=Ps'*Ps;                 // initial information matrix
  ka(i)=10;                        // initial counters
  [Est(i).th, Est(i).cv]=..        // point estimates
  v2thN(Est(i).V/ka(i));           //   of components par.
end
nu=ones(nc,nc);                     // initial pointer statistics       // 11
w1=ones(nc,1)/nc;                   // initial pointer distribution     // 12
al=fnorm(nu,2);                     // pt.est. of pointer par.          // 13

// ESTIMATION
for t=2:nd // TIME LOOP -------------------------------------------- // 14
  yt=y(t);                          // measured output                  // 15

  // proximities of yt from components
  for i=1:nc
    th=Est(i).th;                   // regression coefficients
    cv=Est(i).cv;                   // covariances of components
    m(i)=GaussN(yt,th,cv);          // density of normal component      // 16
  end

  // weighting vector
  Wp=(w1*m').*al; W=Wp/sum(Wp);     // weighting matrix W               // 17
  wp=sum(W,'r')'; w=wp/sum(wp);     // weighting vector w               // 18

  // update of statistics
  Ps=[yt;1];                        // ext. regression vector           // 19
  for i=1:nc                        //                                  // 20
    Est(i).V=Est(i).V+w(i)*Ps*Ps';  // components - matrix V
    ka(i)=ka(i)+w(i);               // components - counter kappa
  end
  nu=nu+W;                          // pointer - matrix nu              // 21

  // point estimates
  for i=1:nc
    [Est(i).th, Est(i).cv]=..       // point estimates                 // 22
    v2thN(Est(i).V/ka(i));          //   of components par.
  end
  al=fnorm(nu,2);                   // point est. of pointer par.       // 23

  w1=w;                             // old weighting vector            // 24
  wt(:,t)=w;                        // saved weighting vector          // 25
end  // END OF TIME LOOP --------------------------------------------
```

```
// RESULTS
[xxx ce]=max(wt,'r');               // estimated active components
wr=sum(c(:)~=ce(:));
printf('\nWrong classifications %d from %d\n',wr,nd)
plot(1:nd,c,'o',1:nd,ce,'.')
```

The procedure *Intro* used at the beginning of the program is an auxiliary file containing the initial settings of a Scilab start. It is presented in Appendix 8.2.1. The functions *v2thN* and *GaussN* can be found in Appendix 8.2.2 and 8.2.3, respectively.

4.1.3   Comments

The first part of the program is devoted to the simulation of a data sample. Data are generated using the one-dimensional static regression models (2.2), whose switching is described by model (3.2).

The program starts with the definition of the length of the simulation (command 1) and the number of components of the mixture model (command 2).

Commands 3–4 define the values of the parameters used for the simulation: the regression coefficients of the i-th component θ_i are denoted by $[Sim(i).th = thS(i)]$ and are

$$\theta_1 = -1, \quad \theta_2 = 1, \quad \theta_3 = 5.$$

Noise variances of the i-th component r_i are $[Sim(i).cv = 0.1]$. They are all 0.1.

Command 5 defines the transition table α corresponding to the pointer model (3.2). It is

$$\alpha = \begin{bmatrix} 0.8 & 0.1 & 0.1 \\ 0.1 & 0.2 & 0.7 \\ 0.3 & 0.1 & 0.6 \end{bmatrix}.$$

Command 6 sets the initial value of the pointer to one: $c_1 = 1$. This is the last command of the initialization of the simulation.

After this, the time loop of the simulation begins. The simulation is performed by commands 7 and 8. First, the value of the pointer c_t is simulated by command 7 and then the output y_t is generated. A more detailed explanation of the pointer value generation can be found in Appendix 8.1.1. The regression models are simulated for each component using the above parameters. After finishing the simulation time loop, the data sample is generated and prepared for the purposes of estimation.

The estimation is initialized by commands 9, 10, and 11. Here, initial values of component parameters are set. According to them, the statistics are constructed (commands 9 and 10) using the function *v2thN*, which provides the computations (2.10). Then command 11 performs the initialization of the statistics of the pointer model.

After this initialization, the time loop of the recursive estimation of the mixture model is presented. It consists of several parts: (i) the computation of proximities of individual components to the actually measured data item; (ii) the construction of the weighting matrix W_t and the weighting vector w_t; (iii) the update of statistics of both component and pointer models using the weights W_t and w_t; (iv) the computation of the point estimates of all parameters from the updated statistics. Comments to these steps are given below.

(i) The proximities are values obtained by the substitution of measured data and previous parameter point estimates into pdfs of estimated components. The closer the measured data item is to the value, which would have been predicted from this component, the larger the computed value. In the program, the simplified version is used (see Appendix 7.9.3). The realization is performed by command 16.

(ii) The weighting matrix W_t and the vector w_t are computed by commands 17 and 18 according to (4.6) and (4.7).

(iii) Command 19 defines the extended regression vector (not forgetting that the example deals with a static regression model), loop 20 updates the statistics of components according to (4.2), (4.3), and command 21 represents the update of the pointer statistics (4.4).

(iv) Command 22 makes the computation of the point estimation for components according to (2.10), while command 23 does it for the pointer via (2.19).

Command 24 saves the value of the weighting vector w_t to be used for the next step of the recursion. Command 25 stores w_t for the final plotting as a result.

At the end of the program, the results are printed and plotted.

4.2   Categorical Components

This section deals with a mixture of m_c categorical components (2.13). Here the general pdf (3.1) used for components in Sect. 3.1 is given by

$$\underbrace{f\left(y_t|\psi_t, \Theta, c_t = i\right)}_{(3.1)} = \underbrace{f\left(y_t = k|\psi_t = l, \beta_i\right)}_{(2.13)}, \ \forall i \in c^*, \tag{4.8}$$

where i corresponds to the i-th component and unlike to (2.13), $k \in y^*$, $l \in \psi^*$ denote values from sets of possible values of discrete variables y_t and ψ_t. Here, to distinguish formally parameters of the categorical components and also for the sake of simplicity, parameters of each i-th component are denoted by β_i. It means that here $\Theta = \{\beta_i\}_{i=1}^{m_c}$, where each β_i is the matrix containing transition probabilities $(\beta_{k|l})_i$ $\forall k \in y^*$ and $\forall l \in \psi^*$ and statement (2.14) holds. Switching the activity of the categorical components is described by the dynamic pointer model (3.2).

Statistics update

The update of statistics of the categorical model (2.13) is performed according to (2.18). Similarly, as in the previous case of regression components, after approximating the Kronecker delta function by the weight w_t, the difference of the updates is only in weighing the data added to the statistics [31]

$$(\eta_{k|l})_{i;t} = (\eta_{k|l})_{i;t-1} + w_{i;t}, \quad \forall k \in y^*, \ \forall l \in \psi^*, \forall i \in c^*, \tag{4.9}$$

performed for each component, where $\eta_{;t}$ denotes the statistics of the i-th component. The statistics of the pointer model is done in a similar way as with the mixture of regression components using (4.4).

Point estimates

Point estimates of parameters of individual components are given by normalizing the statistics $\eta_{i;t}$ according to (2.19). The point estimate of parameter α is obtained similarly.

Prediction

The output prediction within each component can be given by the model with the substituted past data and point estimates of parameters via (2.20).

Comparing the above relations with those from Sect. 4.1, it can be seen that the presented approach keeps its unified form. It differs only in the updates of the component statistics. Let's see the algorithm for a time step of the recursive mixture estimation in the case of categorical components.

4.2.1 Algorithm

One step of the recursive estimation at time t

1. Let's have point estimates of parameters $\left\{\hat{\beta}_{i;t-1}\right\}_{i=1}^{m_c}$ and $\hat{\alpha}_{t-1}$, which are either obtained from the previous step of the estimation according to (2.19) or chosen as the prior ones.
2. Let's have the m_c-dimensional weighting vector w_{t-1}, which is the estimated distribution of the pointer c_{t-1}, i.e., $w_{i;t-1} = f(c_{t-1} = i|d(t-1))$, obtained from the previous step of the estimation or chosen as the initial one.
3. Measure a new data item $d_t = \{y_t, u_t\}$, where y_t and u_t are discrete variables.
4. For each component $i \in c^*$ take the current data item and the previous point of estimate of β and evaluate the proximity of the current data to each i-th component

$$m_i = f\left(y_t = k|\psi_t = l, \hat{\beta}_{i;t-1}\right),$$

where the regression vector ψ_t is coded according to Sect. 2.2.

5. Construct the weight matrix W_t, which contains the pdfs $W_{j,i;t}$ (joint for c_t and c_{t-1})

$$W_t \propto \left(w_{t-1}m'\right) .* \hat{\alpha}_{t-1} =$$

$$= \left\{ \begin{bmatrix} w_{1;t-1} \\ w_{2;t-1} \\ \cdots \\ w_{m_c;t-1} \end{bmatrix} [m_1, m_2, \cdots, m_{m_c}] \right\} .* \begin{bmatrix} \hat{\alpha}_{1|1} & \hat{\alpha}_{2|1} & \cdots & \hat{\alpha}_{m_c|1} \\ \hat{\alpha}_{1|2} & & & \\ \cdots & \cdots & & \cdots \\ \hat{\alpha}_{1|m_c} & \hat{\alpha}_{2|m_c} & \cdots & \hat{\alpha}_{m_c|m_c} \end{bmatrix}, \quad (4.10)$$

where $.*$ is a "dot product" that multiplies the matrices entry by entry.

6. Normalize the matrix W_t from (4.10) so that the overall sum of all its entries is equal to 1.

7. Perform the summation of the normalized matrix W_t over rows and obtain the weighting vector w_t (the distribution of the pointer c_t) with updated entries $w_{i;t}$ $\forall i \in c^*$ according to (3.19) as follows:

$$f(c_t = i | d(t)) = w_{i;t} \propto \sum_{j=1}^{m_c} W_{j,i;t}, \quad (4.11)$$

which gives probabilities of the activity of individual components at the current time instant t.

8. Update the statistics of each i-th component according to (4.9) and the statistics of the pointer using (4.4).

9. Recompute the point estimates of parameters $\hat{\beta}_{i;t}$ of each i-th component according to (2.19).

10. Recompute the point estimate $\hat{\alpha}_t$ of the pointer model with the help of (2.19).

11. Use $\hat{\beta}_{i;t}$, $\hat{\alpha}_t$, and w_t for the next step of the recursive estimation as the prior ones.

A simple program with the algorithm implementation in Scilab follows.

4.2.2 Simple Program

```
// Estimation of dynamic mixture with categorical components
exec("Intro.sce",-1), rand('seed',139737), disp 'Running ...'

nd=1000;                          // number of data               //  1
nc=3;                             // number of components         //  2

// INITIALIZATION of simulation ------------------------------------
Sim(1).be=[.96 .04];              // param. of 1st component
Sim(2).be=[.02 .98];              // param. of 2nd component       //  3
Sim(3).be=[.35 .65];              // param. of 3nd component
alS=[.8 .1 .1;.1 .2 .7;.3 .1 .6]; // parameters of the pointer     //  5

// SIMULATION
```

```
c(1)=1;                              // initial component label        // 6
for t=2:nd
  rnd=rand(1,2,'unif');
  c(t)=(rnd(1)>alS(c(t-1),1))+1;   // pointer value                  // 7
  y(t)=(rnd(2)>Sim(c(t)).be(1))+1; // output value                   // 8
end

// INITIALIZATION of estimation --------------------------------------
Est(1).et=[20 1]*10;                 // initial comp. statistics
Est(2).et=[1 20]*10;                 // initial comp. statistics       // 9
Est(3).et=[10 10]*10;                // initial comp. statistics
nu=ones(nc,nc);                      // initial pointer statistics     // 10
w1=fnorm(ones(nc,1));                // initial weighting vector       // 11

for i=1:nc
  Est(i).be=Est(i).et/sum(Est(i).et); // comp. point estimates        // 12
end
al=fnorm(nu,2);                      // point estimates pointer        // 13
                                     //   (normalization)
// ESTIMATION
for t=2:nd // TIME LOOP -------------------------------------------- // 14
  yt=y(t);                           // measured output                // 15

  // proximities of yt from components
  for i=1:nc
    m(i)=Est(i).be(yt);                                               // 16
  end
  // weighting vector
  Wp=(w1*m).*al; W=Wp/sum(Wp);     // weighting matrix W             // 17
  wp=sum(W,'r')'; w=wp/sum(wp);     // weighting vector w             // 18

  // update of statistics
  for i=1:nc                         //                                // 19
    Est(i).et(yt)=Est(i).et(yt)+w(i);
  end
  nu=nu+W;                           // pointer - matrix nu            // 21

  // point estimates
  for i=1:nc                                                          // 22
    Est(i).be=Est(i).et/sum(Est(i).et);
  end
  al=fnorm(nu,2);                    // point est. of pointer par.     // 23

  w1=w;                              // old weighting vector           // 24
  wt(:,t)=w;                         // saved weighting vector         // 25
end   // END OF TIME LOOP -------------------------------------------

// RESULTS
[xxx,ce]=max(wt,'r'); wr=sum(c(:)~=ce(:));
printf('Wrong classifications %d from %d\n',wr,nd)
plot(1:nd,c,'o',1:nd,ce,'.')
```

4.2.3 Comments

Here again the first part of the program is devoted to the simulation of a data sample with the help of the one-dimensional static categorical models (2.13) switching according to the dynamic pointer model (3.2).

The program defines the length of simulation by command 1 and the number of components by command 2.

Commands denoted by 3 assign values of the component parameters β_i, which are here denoted by $[Sim(i).be]$, as

$$\beta_1 = [0.96, 0.04], \quad \beta_2 = [0.02, 0.98], \quad \beta_3 = [0.35, 0.65].$$

Command 5 defines the transition matrix α of the pointer model (3.2) as follows:

$$\alpha = \begin{bmatrix} 0.8 \ 0.1 \ 0.1 \\ 0.1 \ 0.2 \ 0.7 \\ 0.3 \ 0.1 \ 0.6 \end{bmatrix}.$$

Command 6 sets the initial value of the pointer to one: $c_1 = 1$.

The simulation is performed by commands 7 and 8. First, the value of the pointer c_t is simulated in command 7 and then the output y_t is generated (for the generation of the pointer value see Appendix 8.1.1). The output is generated in the same way as the pointer (the types of the models are similar—both with the categorical distribution) but with the parameter β instead of α.

After the simulation task, the estimation is initialized by commands 9, 10, and 11. Here, the initial values of component parameters are set through values of the initial statistics (command 9) and the same for pointer (command 10). Commands 11–13 compute the point estimates of the initial parameters.

After this initialization, the time loop of the recursive mixture estimation is given. It consists of the same parts as in the case of normal regression components: (i) the computation of proximities of individual components to the actual data item measured; (ii) the construction of the weighting matrix W_t and the weighting vector w_t; (iii) the update of statistics of both the component and the pointer models using the weights W_t and w_t; (iv) the computation of point estimates of all parameters using the updated statistics. The comments to these steps are as follows:

  (i) Here the proximities are represented by the point estimates of the component parameters β_i, i.e., by their normalization. It is performed by command 16.
 (ii) The weighting matrix W_t and the vector w_t are computed by commands 17 and 18 according to (4.6) and (4.7).
(iii) Commands from 19 to 21 perform the updates of statistics according to (4.9) and (4.4).

(iv) The recomputation of the point estimates is performed according to (2.19), which is the same as in the case of computating the proximities. However, it is performed after the final update of the statistics with the weighting matrix W_t and the vector w_t by commands 22 and 23.

Command 24 saves the value of the weighting vector w_t to be used for the next step of the recursion. Command 25 stores w_t for the final plotting as a result.

At the end of the program, the results are printed and plotted.

4.3 State-Space Components

This section focuses on the mixture of m_c state-space components (2.21)–(2.22). Here the general pdf (3.1) used for components in Sect. 3.1 is formed by pdfs

$$f(x_t|x_{t-1}, u_t, c_t = i) = N(M_i x_{t-1} + N_i u_t + F_i, \, \omega_t), \qquad (4.12)$$

$$f(y_t|x_t, u_t, c_t = i) = N(A_i x_t + B_i u_t + G_i, \, v_t), \qquad (4.13)$$

existing $\forall i \in c^*$, where all involved matrices of parameters supposed to be known exist for each component, and ω_t along with v_t are normal white noises with zero expectations and covariance matrices R_{ω_i} and R_{v_i} supposed to be known. Switching the activity of the state-space components is modeled by (3.2).

State estimation

The general algorithm from Sect. 3.2.4 is still valid and needs only some specifications concerning the state-space model. Here, the Kalman filter according to Sect. 2.3.1 is performed for each component and provides the estimated state distribution in the form

$$N(\hat{x}_{t|t_i}, R_{t|t_i}), \qquad (4.14)$$

where $\hat{x}_{t|t_i}$ is the estimated state expectation obtained for the i-th component and $R_{t|t_i}$ is the estimated state covariance matrix for the i-th component, $\forall i \in c^*$ and the output prediction

$$N(\hat{y}_{t_i}, R_{y_i}), \qquad (4.15)$$

where $\hat{y}_{t_i}$ is the predicted output expectation and R_{y_i} is its covariance matrix, both corresponding to the i-th component.

Again the algorithm of the recursive estimation for the mixture of state-space components is presented below [33] based on [8].

4.3.1 Algorithm

One step of the recursive estimation at time t

1. Let's have the estimated normal state distribution $N(\hat{x}_{t-1|t-1}, R_{t-1|t-1})$ and the point estimate $\hat{\alpha}_{t-1}$ of the pointer model, both either obtained from the previous step of the state estimation or chosen as the prior ones.
2. Let's have the m_c-dimensional weighting vector w_{t-1}, which is the estimated distribution of the pointer c_{t-1}, i.e., $w_{i;t-1} = f(c_{t-1} = i|d(t-1))$, obtained from the previous step of the estimation or chosen as the initial one.
3. Let's have all parameters M_i, N_i, F_i, A_i, B_i, G_i, R_{ω_i}, R_{v_i} $\forall i \in c^*$.
4. Measure a new data item $d_t = \{y_t, u_t\}$.
5. For each component $i \in c^*$ run the Kalman filter (2.29)–(2.35) and obtain (4.14) and (4.15).
6. For each component $i \in c^*$ take the current data item and the output prediction (4.15) and evaluate the proximity of the current data to each i-th component

$$m_i = (2\pi)^{-k_y/2}|R_{y_i}|^{-1/2} \exp\left\{-\frac{1}{2}\left(y_t - \hat{y}_{ti}\right)' R_{y_i}^{-1} \left(y_t - \hat{y}_{ti}\right)\right\}. \qquad (4.16)$$

7. Construct the weight matrix W_t, which contains the pdfs $W_{j,i;t}$ (joint for c_t and c_{t-1})

$$W_t \propto \left(w_{t-1}m'\right).* \hat{\alpha}_{t-1} =$$

$$= \left\{\begin{bmatrix} w_{1;t-1} \\ w_{2;t-1} \\ \cdots \\ w_{m_c;t-1} \end{bmatrix} [m_1, m_2, \cdots, m_{m_c}]\right\}.* \begin{bmatrix} \hat{\alpha}_{1|1} & \hat{\alpha}_{2|1} & \cdots & \hat{\alpha}_{m_c|1} \\ \hat{\alpha}_{1|2} & & & \\ \cdots & \cdots & & \cdots \\ \hat{\alpha}_{1|m_c} & \hat{\alpha}_{2|m_c} & \cdots & \hat{\alpha}_{m_c|m_c} \end{bmatrix}, \qquad (4.17)$$

where $.*$ is a "dot product" that multiplies the matrices entry by entry.
8. Normalize the matrix W_t from (4.17) so that the overall sum of all its entries is equal to 1.
9. Perform the summation of the normalized matrix W_t over rows and obtain the weighting vector w_t (the distribution of the pointer c_t) with updated entries $w_{i;t}$ $\forall i \in c^*$ according to (3.19) as follows:

$$f(c_t = i|d(t)) = w_{i;t} \propto \sum_{j=1}^{m_c} W_{j,i;t}, \qquad (4.18)$$

which gives probabilities of the activity of individual components at the current time instant t.
10. Update the statistics of the pointer using (4.4).

11. Obtain the normal pdf approximating the state estimate pdfs from individual components to minimize the Kerridge inaccuracy [41] (see the derivations of the used approximation, e.g., in [33]). Using the results of m_c Kalman filters from step 5 and the weighting vector w_t, the resulted approximated pdf is obtained by

$$\hat{x}_{t|t} = \sum_{i=1}^{m_c} w_{i;t}\hat{x}_{t|t_i}, \tag{4.19}$$

$$R_{t|t} = \sum_{i=1}^{m_c} \left[w_{i;t}R_{t|t_i} + \left(\hat{x}_{t|t} - \hat{x}_{t|t_i}\right)\left(\hat{x}_{t|t} - \hat{x}_{t|t_i}\right)' \right]. \tag{4.20}$$

12. Recompute the point estimate $\hat{\alpha}_t$ of the pointer model with the help of (2.19).
13. Use $\hat{x}_{t|t}$, $R_{t|t}$, $\hat{\alpha}_t$, and w_t for the next step of the recursive estimation as the prior ones.

A simple Scilab program demonstrating the algorithm implementation is presented below.

4.3.2 Simple Program

```
// Estimation of dynamic mixture with state-space components
exec("Intro.sce",-1),rand('seed',139737), disp 'Running ...'

nd=1000;                        // number of data              //  1
nc=3;                           // number of components        //  2

// INITIALIZATION of simulation ------------------------------------
// parameters of components for simulation
m=[.8 .2 .6];                   // pars of the state simulation //  3
k=[ 1 -1  0];                   //    x(t)=mi*x(t-1)+ki+wi,
a=[ 1  1  1];                   //    y(t)=ai*x(t)+vi
Rw=.1; Rv=.1;                   // model noise variances        //  4
// parameters of the pointer for simulation
alS=[.8 .1 .1;.1 .2 .7;.3 .1 .6]; // parameters of the pointer //  5
// initial values
c(1)=1; x(1)=0;                 // initial pointer and state    //  6

// SIMULATION
for t=2:nd
  ru=rand(1,1,'unif');
  c(t)=(ru>alS(c(t-1),1))+1;    // pointer                      //  7
  rn=rand(2,1,'norm');
  x(t)=m(c(t))*x(t-1)+k(c(t))+.1*rn(1); // state value          //  8
  y(t)=a(c(t))*x(t)+.1*rn(2);   // output value
end

// INITIALIZATION of estimation ------------------------------------
Rx=100;                         // variance of state estimate   //  9
xx=0;                           // initial state estimate
nu=ones(nc,nc);                 // initial pointer statistics   // 10
```

```
w1=fnorm(ones(nc,1));              // initial weighting vector           // 11
al=fnorm(nu,2);                    // point estimates - pointer          // 12

// ESTIMATION
for t=2:nd // TIME LOOP ------------------------------------------------- // 14
  yt=y(t);                         // measured output                    // 15

  // proximities of yt from components
  m=zeros(nc,1);
  for i=1:nc
    // Kalman filter for the c-th component
    KF(i)=KFilt(xx,y(t),m(i),a(i),k(i),Rw,Rv,Rx);
    m(i)=GaussN(y(t),KF(i).yp,KF(i).Ry);                                 // 16
  end

  // weighting matrix and vector
  Wp=(w1*m').*al;  W=Wp/sum(Wp);   // weighting matrix W                 // 17
  wp=sum(W,'r')';  w=wp/sum(wp);   // weighting vector w                 // 18

  // update of statistics
  xx=0;  Rx=0;
  for i=1:nc
    xx=xx+w(i)*KF(i).xt;           // merging of means                   // 19
    Rx=Rx+w(i)*(KF(i).Rx+(xx-KF(i).xt)*(xx-KF(i).xt)'); //              // 20
  end                              // merging of variances

  nu=nu+W;                         // pointer statistics update          // 21
  al=fnorm(nu,2);                  // point estimatre of alpha           // 23

  w1=w;                            // old weighting vector               // 24
  wt(:,t)=w;                       // saved weighting vector             // 25
  xh(:,t)=xx;                      // saved state estimate               // 26
end  // END OF TIME LOOP -----------------------------------------------

// RESULTS
[xxx ce]=max(wt,'r');  wr=sum(c(:)~=ce(:));
printf('Wrong classifications %d from %d\n',wr,nd)
plot(1:nd,c,'o',1:nd,ce,'.')
```

The function *KFilt* used inside the program can be found in Appendix 8.2.5.

4.3.3 Comments

In the beginning of the program a data sample is generated by scalar state-space models

$$x_t = M_i x_{t-1} + k_i + \omega_t \tag{4.21}$$

$$y_t = A_i x_t + v_t, \tag{4.22}$$

whose switching is described by the model (3.2).

Command 1 defines the length of simulation and command 2 sets the number of mixture components. Commands 3–4 define parameters M_i, which are grouped in the vector m so that

$$M_1 = 0.8, \quad M_2 = 0.2, \quad M_3 = 0.6.$$

Similarly, parameters k_i and A_i are defined for each $i \in \{1, 2, 3\}$. Noise variances R_v and R_ω both with the values 0.1 are the same for all 3 components.

Command 5 defines the transition matrix α of the pointer model

$$\alpha = \begin{bmatrix} 0.8 & 0.1 & 0.1 \\ 0.1 & 0.2 & 0.7 \\ 0.3 & 0.1 & 0.6 \end{bmatrix}.$$

Command 6 sets the initial value of the pointer $c_1 = 1$ and of the state $x_0 = 0$.

The simulation is performed by commands 7 and 8. First, the value of the pointer c_t is simulated in command 7 and then the state x_t and the output y_t are generated. For the generation of the pointer value see Appendix 8.1.1. The state and output are generated by the state-space model (4.21) and (4.22).

After the simulation task, the estimation is initialized by commands 9–12. Here, the initial covariance matrix of the state estimate (command 9) and initial value of the state estimate (command 10) is defined. Command 11 performs the initialization of the statistics of the pointer model. Command 12 computes the point estimate of the pointer parameter.

After this initialization, the time loop of the recursive estimation runs. It consists of several similar parts as the previous algorithms. However, some of their parts are hidden in the Kalman filter. Remind that the parts are: (i) the computation of proximities; (ii) the construction of the weights; (iii) the updates of statistics and (iv) the computation of the point estimates.

In the time loop, first, the new data item is measured (command 15). Then, for all components the Kalman filter is called with the same initial estimates and one step of the estimation is performed for all components. Also, for each component the predictive pdf is defined and its value for the measured data is computed (command 16). The value is proportional to the proximity of the measured data item to the components. Commands 17 and 18 compute the weighting matrix W_t and the vector w_t in a similar way as for algorithms from Sects. 4.1 and 4.2. Commands 19 and 20 merge the estimates from all components to the single approximated pdf (the normal one with the merged characteristics). Commands 21 and 23 compute the update of the statistics of the pointer model and the corresponding point estimates.

Commands 24–26 save weights as well as the state estimates for the next step and for plotting.

At the end of the program, the results are printed and plotted.

Chapter 5
Program Codes

The previous chapters described the estimation approach to dynamic mixtures of different types of components and demonstrated simple program codes. However, as it has been mentioned before, the discussed algorithms keep the unified form. In order to stress this important feature, this chapter represents another version of the Scilab implementation of the described algorithms, which includes a single main program with options of selecting the specific types of components as necessary. The structure of the programs divides the mixture estimation task into two subtasks, namely:

- Estimation of the pointer model, which has a common code for all programs;
- Estimation of components, depending upon the actual type of components considered. The program codes are algorithmically the same. However, they differ in choosing the model pdfs, statistics, the way of their updates and computing the point estimates. That is why individual subroutines for each type of components are written separately and called as sub-programs.

5.1 Main Program

The code for the main program is presented below. For better presentation, useful comments to the main program structure are given after the code listing.

```
// Mixture estimation for all model types
//  - components: tMix=1   normal regression model
//                tMix=2   categorical model
//                tMix=3   normal state-space model
//  - dynamic pointer model
// ------------------------------------------------------------------
exec("Intro.sce",-1)

// CHOICE OF COMPONENT TYPE
```

© The Author(s) 2017
I. Nagy and E. Suzdaleva, *Algorithms and Programs of Dynamic Mixture Estimation*, SpringerBriefs in Statistics, DOI 10.1007/978-3-319-64671-8_5

```
tMix=2;                                // 1=regr, 2=categ, 3=state

// TASK SPECIFICATION
nd=100;                                // number of data
fi=.99999;                             // exponential forgetting
np=5;                                  // length of prediction

// DATA LOAD  (simulators are E1Sim, E2Sim, E3Sim)
select tMix
  case 1 then                          // - REGRESSION MIXTURE
    load('_data/datRegEx','yt','ct','yi','ths','cvs','als','ncs')
  case 2 then                          // - DISCRETE MIXTURE
    load('_data/datDisEx','yt','ct','yi','als','ps','ncs','kys')
  case 3 then                          // - STATE-SPACE MIXTURE
    load('_data/datStaEx','yt','ct','als','ncs','M','A','F','Rw','Rv')
end
nc=ncs; // no. of components is taken from simulation

// INITIALIZATION -------------------------------------------------
// initialization of the pointer statistics
w=ones(nc,1)/nc;                                   // initial weights
nu=5*eye(nc,nc)+1*rand(nc,nc,'unif');              // statistics
al=fnorm(nu,2);                                    // point estimates

// initialization of the component statistics
select tMix
  case 1 then [Est,ka]=initReg(yi);                // ini.com.stats
  case 2 then mu=initDis(yi,max(yt,'c'));          // ini.com.stats
  case 3 then Rx=eye(2,2)*1e6; xx=zeros(2,1);      // ini. KF stats
end
if tMix==1, EstI=Est; end

// TIME LOOP ------------------------------------------------------
printf('running ...........|\n '),itime=0;
for t=1:nd                                                      // 1
  itime=itime+1; if itime>(nd-1)/20, mprintf('.'), itime=0; end
  // Computation of data-component "proximities"
  select tMix                                                  // 2
    case 1 then mp=predReg(nc,yt(:,t),Est);                    // 3
    case 2 then mp=predDis(nc,yt(:,t),mu);                     // 4
    case 3 then [mp,KF]=predSta(nc,xx,yt(:,t),M,A,F,Rw,Rv,Rx); // 5
  end                                                          // 6
  fp=al'*w;                          // weights for output prediction
  Wp=(w*mp').*al; W=Wp/sum(Wp);      // weighting matrix W       // 7
  wp=sum(W,'r')'; w=wp/sum(wp);      // weighting vector w       // 8
  // above: w   weights - prior/posterior pdf of pointer
  //        mp  are "proximities" of data from components
  //        al  is point estimate of transition table

  // Update of statitics and point estimates of parameters
  // - statistics of the pointer
  nu=nu+W;                           // update of statistics     // 9
  al=fnorm(nu,2);                    // point estimate of alpha  // 10
  // - statistics of components
  select tMix                                                   // 12
    case 1 then [Est,yp,ka]=updtReg(w,yt(:,t),Est,ka,fi,fp);    // 13
    case 2 then [Est,yp,mu]=updtDis(w,yt(:,t),mu);              // 14
    case 3 then [xx,Rx,yp]=updtSta(w,KF);                       // 15
    xxt(:,t)=xx;                     // store                    // 16
```

```
end                                                          // 17
ypt(:,t)=yp;                       // store                  // 18
wt(:,t)=w;                         // store (components)     // 19
wy(:,t+np)=(al')^np*w;             // store (prediction)     // 20

// storing current results
if tMix~=3                                                   // 21
  for i=1:nc                                                 // 22
    Res(i).th(:,t)=Est(i).th(:);   // i-th component center  // 23
  end                              // only for regression model  // 24
end                                                          // 25
end                                                          // 26

// end of TIME LOOP  ----------------------------------------------------
[xxx Ect]=max(wt,'r');             // pt.est. - active comp. // 27
[xxx Ecp]=max(wy,'r');             // pt.est. - predic. comp.// 28

if (max(ct)==max(Ect))
[q,T]=c2c(ct,Ect);
Tct=q(Ect);
if prod(sum(T,1))==0
  disp ' !!! At least one component is not found !!!'
end
end

select tMix
case 1 then exec('Results1.sce',-1)
case 2 then exec('Results2.sce',-1)
case 3 then exec('Results3.sce',-1)
end
```

5.1.1 Comments

The main program includes three key parts as follows.

Task specification, data load and initialization

Here the options for the program launch are set:

- *tMix* selects a type of components;
- *nd* specifies the number of data used for the estimation;
- *nc* is the number of mixture components;
- and *np* specifies the number of steps for the prediction (used for the evaluation of results).

The data load part defines which specific data sample should be chosen depending on the type of components. These data samples are generated separately using the corresponding simulation program either from Sects. 8.1.2, 8.1.3, or 8.1.4. Here the data samples are used only to be loaded.

The initialization deals with setting the initial statistics of the pointer model (common for all mixture models) and the components (specific for each type).

The time loop

The time loop performs a recursive estimation. For the sake of clarity, commands inside it are numbered and explained below.

Commands 2–6 provide the computation of proximities of the data actually measured from the individual components by choosing the actual component type considered. The proximities are stored in the vector mp.

Command 7 constructs and normalizes the matrix W_t as in (4.6) using the previously estimated weighting vector w_{t-1} denoted by w, the vector with obtained proximities mp and previous point estimate of the parameter α denoted by al. Command 8 sums the weighting matrix over rows according to (4.7). Finally, w is the new weighting vector to be used for statistics updates.

These updates are implemented by commands 9–18. Commands 9–10 perform the update of the pointer model statistics denoted by nu and recomputes the point estimate al. Commands 12–17 switch among the types of components to choose the corresponding function, which gives the update of statistics of the components, recomputes the parameter point estimates, and provides the output prediction. The chosen function takes the previous statistics, the weighting vector as well as the current data item as its input arguments and returns the actualized values. Command 18 stores the computed output prediction for the purpose of displaying the obtained results.

Command 19 saves the actual weighting vector. Command 20 computes the np steps ahead pointer prediction as in (3.27) and saves it as well.

Commands 21–26 store the current results.

Results and plots

Command 27 provides the point estimate of the pointer value. Command 28 does the same for the np steps ahead prediction of the pointer value. The auxiliary subroutine $c2c$ checks the results of the estimation. It is given below.

```
function [q,T]=c2c(ct,Ect)
  // cc=c2c(ct,Ect)    permutation of pointer values for a plot
  // ct        simulated pointer
  // Ect       estimated pointer
  // q         order vector for Ect
  // T         transformation matrix
  //
  // USAGE:        ct=q(Ect)
  // set:   [q T]=c2c(ct,Ect);     sim. and estim. pointer
  // plot(1:nd,ct,1:nd,q(Ect))     plotting
  //
  n=min([length(ct),length(Ect)]);
  if max(ct(1:n))~=max(Ect(1:n))
    disp 'ERROR in c2c.sci: Different number of components'
    return
  end

  nc=max(ct);
  T=zeros(nc,nc);
  for t=1:n
    T(ct(t),Ect(t))=T(ct(t),Ect(t))+1;   // transformation matrix
  end
```

```
for i=1:nc
   [xxx,q(i)]=max(T(:,i));                 // order vector
end
endfunction
```

The rest of the programs display and plot the results.

5.2 Subroutines

The subroutines used in the above main program are presented and explained in this section.

5.2.1 Initialization of Estimation

Regression components

The following subroutine performs the initialization of statistics for the parameter estimation for the mixture of regression components, see Sect. 4.1. The initialization uses the initial data denoted by yi, computing their average. This average represents the center of data around which the centers of the initial components should be scattered. Separately for each component, the following initialization steps run:

1. The center of the component is generated by the random shifting of the overall center computed from the initial data sample.
2. Around this newly generated center, a data sample is generated with a "small" covariance matrix. The length of the generated data sample is equal to the length of the original initial data yi.
3. This new data sample is used for the estimation of the created component.
4. The result of the initialization of the component is its updated statistics V, ka and the point estimates of the parameters $theta$ and $covar$.

The subroutine is presented below.

Program

```
function [Est,ka]=initReg(yi)
   // [Est,ka]=initReg(yi)    init. of mixture estimation with
   //                         static 2-dim. regression components
   //                         and dynamic pointer model
   // Est     vector of structures documenting estimation of components
   // ka      vector of data counters
   // yi      init. data (with known pointer)
   //

   nc=max(yi(3,:));
   ni=size(yi,2);
   for i=1:nc
```

```
    Est(i).V=zeros(3,3);       // inf matrix for 2dim static component
    ka(i)=0;
    for t=find(yi(3,:)==i)
      Ps=[yi(1:2,t); 1];
      Est(i).V=Est(i).V+Ps*Ps';
      ka(i)=ka(i)+1;
    end
    [Est(i).th,Est(i).cv]=v2thN(Est(i).V/ka(i),2);

    Est(i).thI=Est(i).th;
    Est(i).cvI=Est(i).cv;
    Est(i).Vi=Est(i).V;
  end
endfunction
```

The auxiliary subroutine *v2thN* can be found in Appendix 8.2.2.

Categorical components

The following subroutine performs the initialization of statistics for estimating the parameters of categorical components, see Sect. 4.2. Here for these purposes, only static components are taken. Thus, the statistic for each of them is a vector with positive numbers. The values of this vector normalized to the sum equal to 1 represent the initial guess of probabilities of components. The magnitudes of the vector values express the belief in the initial guess.

The length of the vector is equal to the number of different values of (here two-dimensional) output y_t. If its entries $y_{1;t} \in \{1, 2\}$ and $y_{2;t} \in \{1, 2\}$, then the different values of y_t are $[1, 1]$, $[1, 2]$, $[2, 1]$ and $[2, 2]$. The number and at the same time the length of the vector statistics is four.

The program works for two-dimensional output only. Its extension to multidimensional data is straightforward.

Program

```
function mu=initDis(yi,b)
  // mu=initDis(yi,b)      initialization of dynamic mixture
  //                       estimation with discrete components
  // mu          component statistics
  // yi          initial data set
  // b           number of values of variables
  //


  nc=max(yi(3,:));
  for i=1:nc
    j=find(yi(3,:)==i);
    y1=yi(1,j);
    y2=yi(2,j);
    tb=tab(y1,y2,b)+.1;
    s=0;
    for k=1:b(1)
      for l=1:b(2)
        s=s+1;
        mu(i).ent(1,s)=tb(k,l);
      end
    end
```

```
  end
endfunction
```

State-space components
The mixture estimation with state-space components does not require any specific initialization subroutines in the main program.

5.2.2 Computation of Proximities

Regression components
For the mixture of regression components, the proximities of the actual data items to the components are computed using the subroutine *predReg*, which is used in the main program, see Sect. 5.1.

The normal distribution (representing the model) is known to fall very rapidly to zero from its expectation. So if the data items inserted into the model are far enough from all components, a zero vector is obtained and it is not possible to distinguish which of the "zeros" obtained is the "biggest one'.' That is why the values of the model are generated as logarithms and it is necessary to convert them to "real" numbers (not logarithms) and normalize. This is done by:

1. Normalize the logarithms by adding them to a constant so that the maximum value is equal to 1. An alternative way is to normalize so that the minimum is equal to one.
2. Take the exponent of all the numbers.
3. Normalize again by dividing the numbers by their sum.

If the proximity of the data is so big that even after this procedure it is not possible to distinguish the numbers, the vector of proximities should be set as uniform. In this case, all items will be the same and equal to $1/nc$, where nc denotes the number of components (in programs only).

The rest of the subroutine is the re-normalization of the probabilities.

Program

```
function mp=predReg(nc,y,Est)
  // mp=predReg(nc,y,theta,covar)  data predic. from reg. components
  // mp      data proximities to components
  // nc      number of comoponents
  // y       data item
  // theta   model regression coefficients
  // covar   model covariance matrices

  mc=zeros(nc,1);
  for i=1:nc
    // proximities of current data y from individual components
    // (from numerical reasons, logarithms are computed, pre-normalized,
    //  returned applying exponential and normalized to probabilities)
    [xxx mc(i)]=GaussN(y,Est(i).th,Est(i).cv);
        // logarithmic"proximities" of data
  end
```

```
mcc=mc-max(mc);                        // pre-normalization
mp=exp(mcc);                           // taking exponentials
if sum(abs(mp))<1e-8
  mp=ones(nc,1);                       // take uniform distr. in need
end
mp=mp/sum(mp);                         // normaliz. to probs
endfunction
```

The subroutine *GaussN* computing the value of the normal multivariate distribution is used inside. It can be found in Appendix 8.2.3.

Categorical components

In the case of categorical components, the proximities of the actual data items to the components are computed using the subroutine *predDis*. It works with the transition table of the categorical model, see Sect. 4.2. The subroutine takes the current data item (denoted by y), receives the number of the row in the transition table from the auxiliary function *psi2row* (which can be found in Appendix 8.2.4) according to this data item and then gets the corresponding probability.

At the end of the subroutine, the point estimates that coincide with the predictions are computed.

Program

```
function mp=predDis(nc,y,mu)
  // mp=predDis(nc,y,mu)  data prediction from discrete components
  // mp       data proximities to components
  // nc       number of components
  // y        data item
  // mu       statistics for model estimation

  j=psi2row(y,[2 2]);                    // number of row in model table

  for c=1:nc
    mup(c).ent=mu(c).ent;
    mup(c).ent(j)=mu(c).ent(j)+1;        // trial updt of stat.
    be(c).ent=fnorm(mup(c).ent,2);       // pt.estim.of par. = predict.
    mp(c)=be(c).ent(j);                  // pred.of c-th comp.
  end
  mp=fnorm(mp);                          // normaliz.of predic.
  mp=mp(:);                              // column vector
endfunction
```

State-space components

In the case of state-space components, the proximities of the actual data items to the components are computed using the subroutine *predSta*. It is based mostly on the Kalman filter (see Sect. 2.3.1 and Appendix 8.2.5) and the function *GaussN* computing the value of the normal multivariate distribution similarly as in Sect. 5.2.2 (see listing the codes and Appendix 8.2.3).

Program

```
function [mp,KF]=predSta(nc,xx,y,M,A,F,Rs,Rv,Rx)
  // predSta(xx,y,M,A,F,Rs,Rv,Rx)  data prediction from regression components
  // mp       proximities of data to components
  // KF       structure of results from KFilt
```

```
// xx       old state estimate
// y        data item
// M,A,F    model parameters
// Rs,Rv    model covariances
// Rx       covariance matrix of the state estimate

for c=1:nc
  KF(c)=KFilt(xx,y,M(c),A(c),F(c),Rs(c),Rv(c),Rx); // KF
  mc(c)=GaussN(y,KF(c).yp,KF(c).Ry); // data-comp. proximity
end
if abs(sum(mc))<1e-8, mc=ones(nc,1)/nc; end // if the current data
// item is too far from all components, the proximities are all zero;
// in that case we choose their uniform distribution.
mp=fnorm(mc);                            // normalization
mp=mp(:);                                // column vector
endfunction
```

5.2.3 Update of Component Statistics

Regression components

The statistics update for the mixture of regression components according to Sect. 4.1 is implemented in the following program.

In addition to the update, just for displaying the results, the data prediction is computed here. For its generation, the square root of the covariance matrix is used as the standard deviation. This operation is implemented in the function *uut* (see Appendix 8.2.6), which generates the upper triangular matrix U so that the covariance matrix $C = UU'$.

Program

```
function [Est,yp,ka]=updtReg(w,y,Est,ka,fi,wy,sq)
  // [theta,covar,yp,V,ka]=updtReg(w,y,V,ka,fi) update of regr. statistics
  // theta    model regression coefficients
  // covar    model covariance matrices
  // yp       data prediction
  // V,ka     component statistics
  // w        actual component weights
  // y        data item
  // fi       factor of exponential forgetting

  if argn(2)<7, sq=0; end
  if sq>0, sq=1; end
  ny=max(size(y));
  nc=max(size(w));                       // number of components
  yp=zeros(ny,1);
  // update of statistics for components
  d=[y;1];                               // extended regression vector
  for c=1:nc
    ka(c)=ka(c)+w(c);                    // update of the counter
    Est(c).V=fi*Est(c).V+w(c)*d*d';      // update of information matrix
    [Est(c).th,Est(c).cv]=v2thN(Est(c).V/ka(c),ny); // point estimates
    sqc=uut(Est(c).cv);                  // sqrt of covariance matrix
    yp=yp+wy(c)*(Est(c).th'+sq*sqc*rand(ny,1,'norm')); // mix. data pred.
```

```
    end
    if t<.25*nd; for i=1:nc, Est(i).cv=.01*eye(2,2); end, end
    // this is a trick: until the parameter estimates are not roughly
    //   precise, we fix the covariances to relatively small values.
    //   Only then, they are allowed to evolve according to the data.
endfunction
```

Categorical components

The statistics update for the mixture of categorical components according to Sect. 4.2 is implemented in the program below. It also uses the auxiliary functions *psi2row* similarly as in Sect. 5.2.2 and the backwards transformation function *row2psi*. Both of them are listed in Appendix 8.2.4.

It should be noted that the data prediction is computed by normalizing the statistics table. It is used for graphs only.

Program

```
function [Est,yp,mu]=updtDis(w,y,mu)
    // [theta,yp,mu]=updtDis(w,y,mu) update of discrete statistics
    // theta    model regression coefficients
    // yp       data prediction
    // mu       statistics for model estimation

    [xxx,wm]=max(w);                        // maximum w
    nc=length(w);                           // number of components
    j=psi2row(y,[2 2]);                     // convert from y1,y2 to table row
    // component statistics update
    yy=0;
    for c=1:nc
      mu(c).ent(j)=mu(c).ent(j)+w(c);       // final update
      Est(c).th=fnorm(mu(c).ent,2);         // point estimates
    end
    [xxx yy]=max(Est(wm).th);               // pointer point estimate
    yp=row2psi(yy,[2 2])';                  // conversion to y1,y2

endfunction
```

The function *fnorm* used inside can be found in Appendix 8.2.7.

State-space components

A specific feature of the mixture estimation with state-space components is the approximation of the state pdfs obtained for each component by a single normal pdf, see Sect. 4.3. This approximation which is based on minimizing Kerridge inaccuracy [41] is implemented in the subsequent program.

Program

```
function [xx,Rx,yp]=updtSta(w,KF)
    // [xx,Rx,yp]=updtSta(fc,xi,Ri,yi)  composition of Kalman filters
    // xx       mixture state estimate
    // Rx       mixture covariance estimate
    // yp       mixture data prediction
    // w        actual covariance weights
    // KF       structure of results from KFilt

    n=max(size(KF));             // number of components
```

```
  distKL=0;     // distKL=1 using KL else geometrical w average
  if distKL
    sx=0;
    s1=0;
    s2=0;
    // state estimate
    for i=1:n
      sx=sx+w(i)*KF(i).xt;     // computation of mixture state
    end
    xx=sx;
    // covariance matrix estimate
    for i=1:n
      s1=s1+w(i)*(xx-KF(i).xt)*(xx-KF(i).xt)'; // mix. cov.matrix
      s2=s2+w(i)*KF(i).Rx;                      // KL of N-pdfs
    end
    pointEstRx=1;
    if pointEstRx==0, Rx=s1+s2; else
      [xxx im]=max(w);
      Rx=KF(im).Rx;
    end
  else
    [xx,Rx]=connect(w,KF);
  end
  // prediction computation
  yp=0;
  for i=1:n
    yp=yp+w(i)*KF(i).yp; // computation of mixture prediction
  end
endfunction
```

The function *connect* performing the approximation can be found in Appendix 8.2.8.

5.3 Collection of Programs

The presented programs are commented in as much detail as possible. However, it is clear that in order to completely understand how they work, a reader should try to run them. For this aim, all programs are available online. Moreover, the auxiliary programs which are not necessary but nevertheless very helpful, are listed in Appendix 8.

Theoretical algorithms and its program implementations are tightly bound. Hence, the program can be viewed as an excellent schematic expression of the theoretical algorithm and serve as its description. The programs can be directly used by those who want to perform experiments with them or to use them in their own work.

Chapter 6
Experiments

The aim of this chapter is to show how the algorithms and programs presented in Chaps. 4 and 5 work. The mixture estimation is an extremely difficult task and its functionality heavily depends on the nature of the used data. That is why in each experiment it is necessary to judge the results, taking into account how suitable for the mixture estimation the concrete data are.

The majority of the experiments are done with simulated data. The reason is that the data can be generated precisely in a form suitable for the feature of the estimator, which is to be demonstrated. For instance, one of the important experiments is to compare the estimation with well-separated mixture components and with those which are blended together. Another tested feature is the ability of the estimator to cope with components that are not Gaussian. It is also significant to show how the estimation will work in the case of a different number of components in simulation and estimation (especially with more simulated components than used for estimation), etc.

In any case, a crucial task for the estimation is its initialization. This task suggests preparing the prior component models (their number and the prior statistics for the parameter estimation) and the prior pointer model (which is less important and a kind of uniformity is acceptable).

For the start of the estimation algorithm, it is necessary to set the main components which are well recognizable in the data from the very beginning. If a component is located too far from the area of the data, its weight is always practically zero. It is not updated and it remains an empty one, which means that the modeling is wrong. In the case of the well-separated data components, the estimation can run from a random start and there is a chance to get correct results. But even in this case it is not recommended to rely on it.

This indicates that practically in all cases it is important to do some initial analysis in the data space and to catch, at least approximately, the most important density clusters. For this task some classical algorithms, e.g., K-means [42–44], can be used.

© The Author(s) 2017

I. Nagy and E. Suzdaleva, *Algorithms and Programs of Dynamic Mixture Estimation*, SpringerBriefs in Statistics, DOI 10.1007/978-3-319-64671-8_6

However, the best possibility as it appears is the use of a set of prior already classified data for preliminary learning the algorithm. The prior data sample is very often available in practice. If such a data sample is not available, it can be artificially produced (by experts, simulations, etc.). The initial classification can be deduced by experts directly from the nature of a problem which has been actually solved. A simple example to demonstrate what is meant is as follows: traffic accidents (light or heavy) are analyzed depending on the car speed and the visibility. The data space here is mixed (the speed belongs to positive real numbers, while the visibility can be defined as {"dark", "dawn", "dusk", "cloudy", and "sunshine"}. Now, the task for experts is to define approximately which combinations of values of speed and visibility are most characteristic to heavy accidents and where only light ones are more expectable. Such knowledge is often presented in various safety transportation reports and can be easily used also for the discussed initialization purposes.

6.1 Mixture with Regression Components

Here experiments with normal regression components (see Sect. 4.1) of the dynamic mixture are presented. This is the basic form of a mixture that covers a majority of practical tasks to be encountered. The presented experiments aim at demonstrating the theory from Sect. 4.1, namely

- the pointer estimation, which is expressed in determining the currently active components;
- the parameter estimation;
- and the mixture prediction from Sect. 3.3.

In this chapter the main emphasis is on static regression components. The data prediction produced by the component declared as currently active (or a combination of all components weighted with the predicted probabilities of active components) can be determined "without any noise" as the expectation of the output. Noise can be added with the estimated variance in order to work with the whole predictive pdf. The expectation should in the ideal case coincide with the center of the active component and the prediction simulated from the whole predictive pdf should cover the whole data cluster. In this case, it will indicate the correct estimation from the point of view of both the expectation and the variance.

For a better demonstration, a two-dimensional data sample has been taken which enables a direct visualization of components.

Simulation Five components are simulated using the following regression coefficients entering the static models (4.1):

$$\theta_1 = [2 \ -2]', \ \theta_2 = [1 \ 5]', \ \theta_3 = [8 \ 1]', \ \theta_4 = [5 \ 4]', \ \theta_5 = [-1 \ 2]', \qquad (6.1)$$

which represent centers of the components during the simulation. The used covariance matrices are

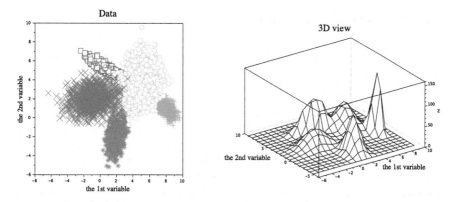

Fig. 6.1 Data with well-separated components. The *left figure* shows clusters of the simulated data obtained by plotting the first entry of the output vector against the second one. The *right* part demonstrates a 3D view of the same data. Here the structure and the shape of the clusters are more clear

$$r_1 = \lambda \begin{bmatrix} 1 & 0.2 \\ 0.2 & 2 \end{bmatrix}, \; r_2 = \lambda \begin{bmatrix} 2 & -0.9 \\ -0.9 & 1 \end{bmatrix}, \; r_3 = \lambda \begin{bmatrix} 1 & -0.3 \\ -0.3 & 1 \end{bmatrix},$$

$$r_4 = \lambda \begin{bmatrix} 2 & 0.2 \\ 0.2 & 3 \end{bmatrix}, \; r_5 = \lambda \begin{bmatrix} 3 & 0.2 \\ 0.2 & 3 \end{bmatrix}, \qquad\qquad (6.2)$$

where the parameter λ sets the noise magnitude different for performed experiments. The generated data sample is visualized in Fig. 6.1.

The pointer model (3.2) for the simulation of switching these components is the (5×5) transition table of nonnegative numbers, which sum to one in rows and whose off-diagonal entries are practically uniform. Diagonal entries are also uniform but approximately ten times higher than the off-diagonal ones. It means that jumps between components are uniform. However, each component stays active for a while before switching to another. This situation is enabled by using the dynamic pointer model, which is defined not only by a vector of probabilities of individual active components but also by the whole transition matrix. Specifically, here the table (3.2) is as follows:

$$
\begin{array}{ccccc}
0.69 & 0.03 & 0.07 & 0.07 & 0.14 \\
0.01 & 0.58 & 0.06 & 0.29 & 0.06 \\
0.05 & 0.02 & 0.62 & 0.19 & 0.12 \\
0.06 & 0.02 & 0.12 & 0.59 & 0.06 \\
0.11 & 0.17 & 0.11 & 0.06 & 0.55
\end{array}
$$

6.1.1   Well-Separated Components

The first experiment deals with the ideal case of the components location, which are well separated and uniformly filled with data with a small noise. Results of the mixture estimation according to Sect. 4.1.1 are as follows.

Pointer estimation Fig. 6.2 confirms the ability of the algorithm to perform the data classification according to active components. It takes the point estimates of the pointer obtained as the indexes of entries of the weighting vector w_t, which reach the maximal probability, and compares them with the simulated values of the pointer. For preliminary learning, five correctly estimated data items for each component have been used. Figure 6.3 demonstrates the evolution of the weights of individual components.

Parameter estimation As it has been already said, the parameter estimation heavily depends on its initialization, especially on the initial component centers (defined by expectations of regression coefficients) and their dispersions (given by covariance matrices). Using the prior classified data sample seems to be the most effective way of the initialization. The estimation with this type of the initialization is shown in Fig. 6.4.

The obtained point estimates are very close to the regression coefficients (6.1) used for simulation. They are

$$\hat{\theta}_{1;t} = [2.07\ -1.72]',\ \hat{\theta}_{2;t} = [1.03\ 4.91]',\ \hat{\theta}_{3;t} = [7.85\ 1.02]',\ \hat{\theta}_{4;t} = [4.58\ 3.88]',\ \hat{\theta}_{5;t} = [-1.5\ 1.77]'. \tag{6.3}$$

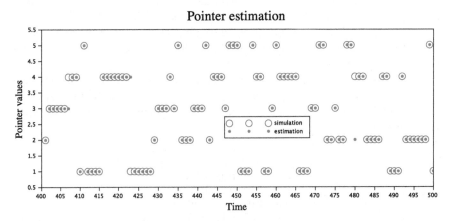

Fig. 6.2 A comparison of simulated and estimated values of the pointer. Simulated values of the pointer are denoted by *circles*, while the estimated ones are denoted by *dots*. At each time instant, they indicate active components of the simulated and the estimated mixture, respectively. Notice the correspondence of the *circles* in relation to the *dots*

Fig. 6.3 The time evolution of weights of five components. Each subplot demonstrates changes of the weights of five components in time during the estimation. Notice that the maximum weights, which are the probability of the activity of components, correspond to the point estimates of the pointer in Fig. 6.2

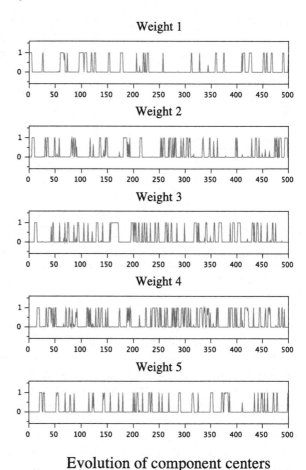

Fig. 6.4 The evolution of parameter point estimates during the estimation with initialization based on prior classified data. The figure presents the time evolution of the point estimates of regression coefficients of each component (i.e., their centers). When a prior pre-classified data sample is used for initialization, the initial settings can be arbitrary. The progress of the estimation is very fast and the convergence to true values (denoted by *stars*) is practically precise

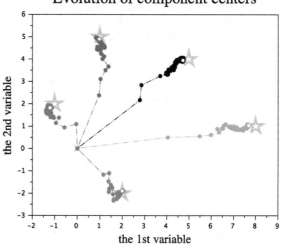

Evolution of component centers with K-means   Evolution with random initialization

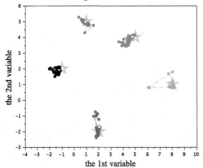 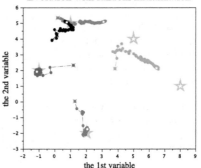

Fig. 6.5 The evolution of parameter point estimates during the estimation with initialization using K-means (*left*) and with the random initialization (*right*) In the *left figure*, the initial values of the estimates are close to the true centers due to the K-means classification. However, K-means frequently joins two components and sets two centers into one data cluster. In the *right figure*, the estimation starts randomly. The initial parameter estimates are more distant from their true positions. However, the convergence is also quick

If the pre-classified data sample is not available, success can be expected only in cases when components are well separated (as it is in this example). In this case, it is necessary to determine the initial component positions according to the data.

Here two possible ways can be recommended: (i) preprocessing the initial data sample by some clustering algorithm, e.g., K-means [44]; (ii) computing the center of the whole initial data sample and distributing the component centers randomly around this center in an average distance given by the sample variance of the data. In any case, it is necessary to start the estimation of the regression coefficients for some 100–200 initial steps, while the component variances are kept fixed with their initial setting and to start the variance estimation only afterward. Otherwise, a single component becomes dominant and increases its dispersion; it covers all component centers, catches all coming data items, and suppresses estimating all other components. Results of the estimation using these initializations methods are demonstrated in Fig. 6.5. Generally, it is advantageous to work with the normalized data sample, for which it holds that the variables have the zero mean values and the sample variance is equal to one. Then the area of the data is delimited well.

The estimated values of the parameter α of the pointer model (3.2) are

$$\begin{array}{ccccc} 0.64 & 0.05 & 0.09 & 0.12 & 0.1 \\ 0.01 & 0.5 & 0.11 & 0.33 & 0.05 \\ 0.09 & 0.03 & 0.56 & 0.17 & 0.15 \\ 0.07 & 0.19 & 0.18 & 0.48 & 0.08 \\ 0.08 & 0.18 & 0.12 & 0.12 & 0.5 \end{array}$$

which in comparison with the simulated values gives a good result.

The estimation in this case, when the simulated components are easy separable, is relatively reliable. Nevertheless, even here wrong classification can occur, which leads to a failure of the whole estimation. Therefore, a kind of validation is always necessary. In the considered simple two-dimensional case, a visual check can be sufficient. In any case, a verification of the results is recommended as it is independent of the dimension of the components. A very simple check of the classification can be done by inspecting the component statistics $\kappa_{i;t}$, $i = 1, 2, \cdots, m_c$ updated according to (4.3). If some entry of this statistics is too small, it means that the corresponding component has practically never been active and therefore it has not been estimated. So, it is superfluous or it has been joined together with some other component. In any event, this fact shows that something is wrong with the estimation.

Prediction The prediction can also serve as a validation of the estimation process. However, for a dimension higher than two (as in this example) the visualization of components is problematic. The most straightforward way is to check all two-dimensional marginals.

According to Sect. 3.3, the prediction can be zero-step ahead, when the current output is estimated before its measuring and multistep ahead, when the future output value is estimated. The prediction can be generated as an expectation of the output with the corresponding predictive pdf. If correct, it coincides with the center of the active component (present or future). Figure 6.6 shows the results with the zero-step prediction.

The multistep prediction, see Sect. 3.3, is rather problematic. The prediction with a mixture with static components can be verbally characterized as follows: make the multistep prediction of the pointer and use it to generate the output prediction (either with the expectations or that with a noise). However, the pointer prediction generally can be very uncertain.

As it was mentioned in the description of the simulation of the data (in the beginning of this chapter), the taken transition table of the pointer model is configured to

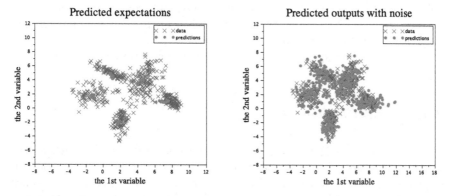

Fig. 6.6 The output prediction in the form of clusters The *left figure* shows the output prediction obtained using expectations. The *right figure* plots the same with a corresponding noise, i.e., using the covariance matrices. Based on the correct estimation, the prediction exhibits adequate results

keep the activity of a component for a while before it jumps to another one. Now, the pointer prediction is governed by a power of this matrix according to (3.27) considered for the necessary number of the prediction steps. Taking the power of such matrix causes falling values on the diagonal and rising for those outside of the diagonal. This means a growth of uncertainty: the jumps can come at anytime and anywhere.

For the prediction, some order in jumping of activities is necessary. For example, Tuesday comes after Monday, a night comes after a day, etc. Such a situation is characterized by a special form of the parameter α. For example, if components (for brevity, three of them) are switched one after another, i.e., 1–2–3–1 – etc., the parameter α has the form (in the deterministic case)

$$\begin{matrix} 0 & 1 & 0 \\ 0 & 0 & 1 \\ 1 & 0 & 0 \end{matrix}$$

and similarly for more components. For a stochastic case, zeros are replaced by small numbers and ones by big numbers, so that the sum of rows will always be one. In a deterministic case, it is possible to predict the activity of components exactly, for stochastic case with a high probability.

The above situation is demonstrated in the following exploration, where the last 100 simulated and predicted pointer values are compared. The three-step-ahead prediction is considered. Results are as follows:

- The experiment with a standard switching the components, i.e., with the parameter α with big diagonal and small non-diagonal entries gives 71 wrong predictions.
- The experiment with a sequential switching with big numbers equal to 0.96 and small ones 0.01 gives seven wrong predictions.

For the more-step prediction, the results in the first case are the same. In the case of sequential switching, the 5-step prediction gives 43 wrong predictions and the 15-step provides 82 wrong predictions, which is approximately the same result as for general switching, when the prediction has no sense.

The results with sequential pointer switching are demonstrated in Fig. 6.7.

6.1.2 Weak Components

The second experiment deals with a difficult situation for the estimator, when some component is badly excited (it is active only rarely). Mostly, the weak component is suppressed during the estimation. However, in the case of using prior pre-classified data for the initialization, where this component is not neglected, it is possible to perform a correct estimation. Here three simulated components are taken. The evolution of the point parameter estimates in this situation is visualized in Fig. 6.8.

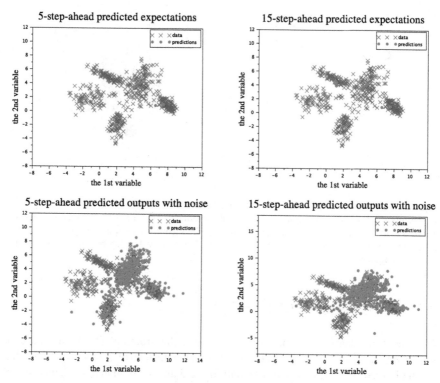

Fig. 6.7 A comparison of simulated and predicted clusters. The *left figures* show the five-step-ahead predicted expectations (*top*) as well as outputs with noise (*bottom*) and compare them with simulated data. It can be seen that the clusters positions are correct. In the *right* part of the figure, the same prediction is plotted for 15 steps ahead. Here, the prediction is incorrect

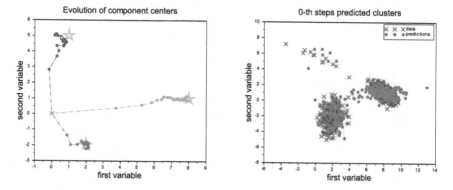

Fig. 6.8 The evolution of the point parameter estimates (*left*) and the output prediction plotted as clusters (*right*) with one weak component. Using a pre-classified initial data sample, the estimation can be successfully accomplished. Here, the second component is weak. Nevertheless, its estimation went smoothly

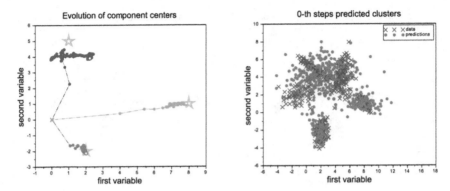

Fig. 6.9 The evolution of the point parameter estimates (*left*) and the output prediction plotted as clusters (*right*) with a different structure of the estimated mixture. The estimation in the *left figure* is evidently successful. In the *right figure* it can be seen that the estimation gives reasonable predictions. Two more remote components are captured separately. The rest of them are covered by a single component

6.1.3   Reduced Number of Components

Obviously, the ideal case for the estimation is when the structure of the model through which the data sample is simulated is identical with that of a model used in the estimation. However, this is possible only if the data are simulated. In reality, there is no simulation model and in addition the measured data items are not precisely normal and influenced by nonlinearities as well as noises. It is quite a common issue in applications. To come a little closer to such a situation, here a mixture of five components is simulated and a mixture with only three components is used for the estimation. The question is how the estimator will work with these three components to reasonably cover the five hills in the data space. The results are shown in Fig. 6.9.

6.1.4   High-Dimensional Output

All examples presented were two-dimensional. The reason is the visualization of the results, which in the two-dimensional case allows for a clear demonstration of the tested properties of the estimation algorithm. However, especially in practical applications, the algorithm will operate on multidimensional data samples. That is why it is necessary to demonstrate such an operation too and at the same time to mention some simple ways of validations of such an estimation process, when direct visualization is not possible.

Here the five-dimensional data sample with three components is simulated. In this case, even the simulated clusters cannot be displayed. The estimation is initialized using the initial pre-classified data (three data items into each component). The

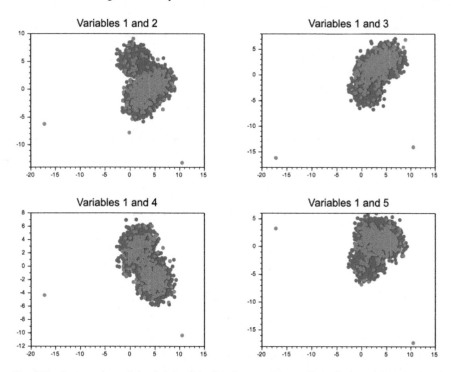

Fig. 6.10 A comparison of simulated and predicted output clusters. Here, the first variable is plotted against the remaining ones. It can be seen that predicted clusters correspond to simulated values, which even in this relatively simple comparison indicates that the estimation was successful

resulting simulated and predicted clusters can be displayed only by plotting each variable against the others. As an example, results for the first variable plotted against the others are demonstrated in Fig. 6.10.

A better way of validating the estimation results of multivariate clusters is to work with the pointer, i.e., to check the classification properties of the estimated mixture. The result is shown in Fig. 6.11.

In a real application, the final values of the statistics $\kappa_{i;t}$ can be useful. This statistics (sometimes called the counter) counts the probabilities of the activity of individual components. Subtracting its initial values from the final statistics allows to obtain the total sum of activities of individual components. If some entry from this total sum is close to zero, there is something wrong with the estimation (too many components, a component has not been used in the estimation, etc.). In this experiment these total sums are

$$115.3, \ \ 81.8, \ \ 302.9$$

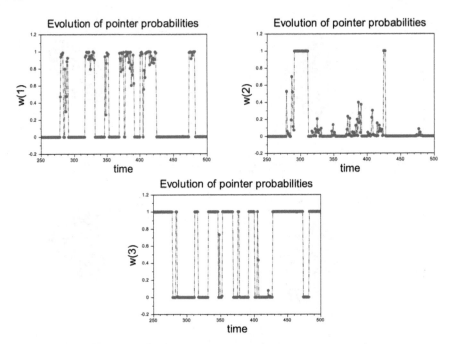

Fig. 6.11 The evolution of probabilities of the activity of individual components. If the decisions of the active component are clear, one of the entries of the weighting vector w_t is close to one and the rest of them are practically zeros. In the case of unclear decisions, the probabilities are near to 0.5. If some component is eliminated, all its probabilities are zero

for the components 1, 2, and 3 respectively. The distribution of the data among the components points out that the estimation was successful, i.e., neither of the components has the zero sum of weights.

6.1.5 Big Noise

For a satisfactory estimation, first of all, the classification (i.e., the active component detection) must be successful. The classification quality depends on the distinguishability of individual components. If the data are noisy and the components are rather overlapping, the estimation of the mixture model is complicated (especially in the beginning). However, with an initial pre-classified data sample, even in such a situation the mixture estimation and the prediction are well solvable. This fact is tested in experiments demonstrated in this section.

Figure 6.12 shows the time evolution of the point estimates of the regression coefficients (centers of individual components). Figure 6.13 provides results of the zero-step prediction of the pointer and the output.

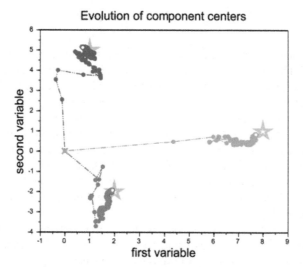

Fig. 6.12 The time evolution of the parameter point estimates of components during the estimation. The figure demonstrates the very quick progress of the correct evolution at the beginning of the estimation given by the initial pre-classified data sample used for the initialization. After a good settlement of the estimates, the situation stabilizes (the estimates stick to true component centers which even in big noise are perceptible). Without the pre-classified data sample such results would be impossible

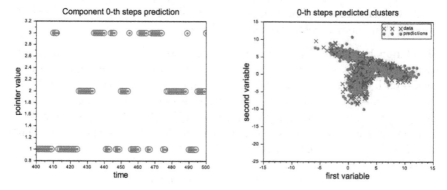

Fig. 6.13 The zero-step prediction of the pointer and the output. The *left plot* demonstrates the simulated and the estimated values of the pointer, expressing the activity of individual components. The *right plot* provides the output prediction visualized in the form of clusters. It can be seen that the clusters are very dense. However, the predicted clusters cover the original data

In this way, the most significant features of Sect. 4.1.1 estimating the mixture of normal regression components are demonstrated. Now, let us see the possibilities of Sect. 4.2.1 dealing with categorical components in the next section.

6.2 Mixture with Categorical Components

This section demonstrates the results of experiments with categorical components, see Sect. 4.2. The aim of the experiments is, similarly as in the previous section, to show the most important features and the successfulness of Sect. 4.2.1 in the tasks of the pointer and parameter estimation and the prediction.

Obviously, the mixture of categorical components has different properties than that of regression models discussed in the previous section, where the uncertainty of a model is given by the magnitude (the variance) of its noise. With a categorical model, the uncertainty is given by entries of the transition table, i.e., by the stationary probabilities of individual values of the variable. If the probabilities are close to 0.5, the model is absolutely uncertain. The probabilities near 0 and 1 indicate a deterministic model.

Another aspect of a categorical model is that it is more difficult for estimation than its continuous counterpart. For example, for the estimation of a constant k of a static model $y_t = k + e_t$ it is enough to have only few values of y_t in order to estimate the expectation of y_t as an average of measured data, which is known not to be data demanding. With small uncertainty, only one value gives a reasonable estimate. For the binary categorical model (e.g., tossing a coin), the estimation of only one value makes no sense. To obtain a reasonable estimate of the probabilities of the individual sides of a coin, one should have many more values. To verify that a possibly damaged coin still provides regular values, hundreds of measurements are required.

On the other side, categorical models or their mixtures can give satisfactory results even for continuous variables properly discretized.

Simulation For simplicity, a simple static categorical model (4.8) is considered (without the regression vector), where the discrete output y_t is the two-dimensional vector. Each entry has two possible values 1 and 2, which means $y_{1;t} \in \{1, 2\}$ and $y_{2;t} \in \{1, 2\}$. According to Sect. 2.2, the dimension of the output vector is reduced to a single variable by coding $[y_1, y_2] = [1, 1] \rightarrow 1, [1, 2] \rightarrow 2, [2, 1] \rightarrow 3, [2, 2] \rightarrow 4$. Then the parameter of each component is the four-dimensional vector. Three components of discrete data are generated by

$$\beta_1 = [0.9, \ 0.03, \ 0.05, \ 0.02]$$
$$\beta_2 = [0.02, \ 0.9, \ 0.02, \ .06]$$
$$\beta_3 = [0.01, \ 0.01, \ 0.54, \ 0.44]. \tag{6.4}$$

The pointer model of their switching is as follows:

$$\alpha = \begin{bmatrix} 0.7, & 0.1, & 0.2 \\ 0.15, & 0.8, & 0.05 \\ 0.1, & 0.1, & 0.8 \end{bmatrix}. \tag{6.5}$$

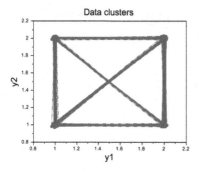
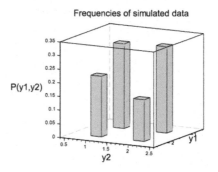

Fig. 6.14 The simulated data for discrete mixture estimation. The *left part* of the figure demonstrates the simulated data in a jiggled form. Corners of the rectangle are jiggled simulated data points. To also show the transitions of the points, their connecting *lines* are shown too. The intensity of the *color* shows the density of points or transitions. The *right part* of the figure shows the densities of individual simulated points

It is problematic to show the simulated data sample because the data forms only a finite number of points in the data space. One possibility is to jiggle the points by adding a very small noise to them (e.g., as used in the software system Weka [46], see http://www.cs.waikato.ac.nz/ml/weka/). Another possibility is to show frequencies of data occurrence on a bar graph. The data sample is shown in Fig. 6.14.

Results Here the results of experiments with the above-simulated mixture are presented. Figure 6.15 presents the time evolution of the point parameter estimates. Figure 6.16 shows results of the pointer estimation and the zero-step output prediction. Figure 6.17 demonstrates the switching of activities of components by plotting the weights of each component in time.

The results demonstrated for the mixture of categorical components are not so numerous as in the case of regression components. They are shortened in order not to bore a reader with the details of the estimation and the great number of figures. However, the possibility of using the estimation algorithm for discrete data and its successfulness are shown. Now, the results of experiments with state-space components can be demonstrated.

6.3   Mixture with State-Space Components

This section is devoted to experiments with a mixture of state-space components, see Sect. 4.3. State-space components represent a specific case of the dynamic mixture estimation, which supposes parameters to be known and estimates the unobserved state. Similarly, the experiments aim at the key features of Sect. 4.3.1 in solving the tasks of the pointer and parameter estimation as well as the prediction.

Simulation Three state-space components (4.12) switching according to the pointer model

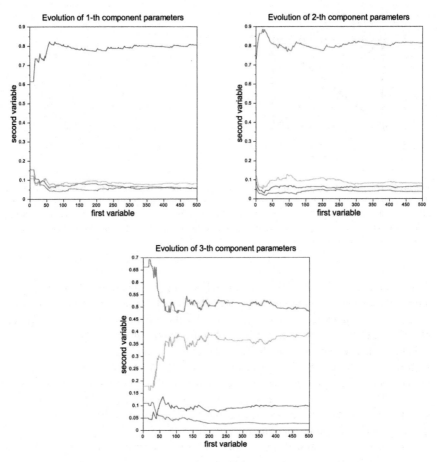

Fig. 6.15 The time evolution of parameter point estimates during the estimation. The figure shows how the point estimates of components change during the estimation. The stabilized values (for the time equal to 500) should be compared with the simulated parameters (6.4). It can be seen that after the initial search, satisfactory values are found and the estimation stabilizes

$$
\alpha = \begin{bmatrix} 0.95, & 0.025, & 0.025 \\ 0.01, & 0.98, & 0.01 \\ 0.02, & 0.02, & 0.96 \end{bmatrix} \tag{6.6}
$$

are simulated with the two-dimensional state x_t and the output y_t, without the control input u_t and the constant G_i, $i \in \{1, 2, 3\}$. The parameters corresponding to (4.12) are

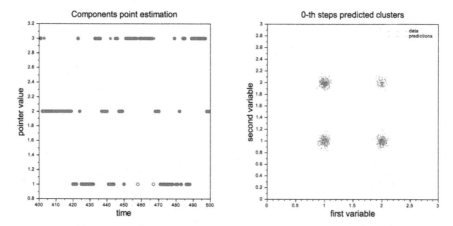

Fig. 6.16 Active component estimation. The *left plot* compares the simulated and the estimated values of the pointer, expressing switching the components. The figure shows an excellent agreement. From the second part of the estimation, i.e., from 250 data items, only 21 wrong classifications were obtained. The *right plot* shows the predicted outputs as clusters and compares them with simulated ones. It can be seen that the estimates satisfactorily cover the data clusters

Fig. 6.17 The weights of individual components. The figure proves that all components are used as active. It means that the whole structure of the mixture model is well proposed, i.e., specific components in the data are found and each of them is modeled by the corresponding model

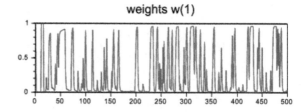

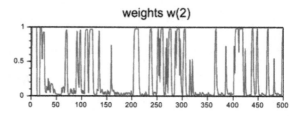

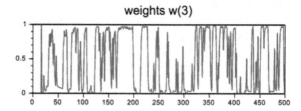

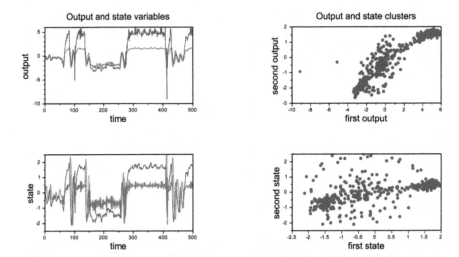

Fig. 6.18 Simulated data. The *left plot* shows the output and the state (both two-dimensional), where switching the components is clearly visible. The *right plot* demonstrates the output and the state in the form of clusters. The clusters are not very separated, so the classification is not trivial

$$M_1 = \begin{bmatrix} 0.83, & -0.3 \\ 0.5, & 0.91 \end{bmatrix}, \quad F_1 = \begin{bmatrix} -1 \\ 2 \end{bmatrix}, \quad A_1 = \begin{bmatrix} 0.9, & 0.4 \\ 1, & -0.4 \end{bmatrix}$$

$$M_2 = \begin{bmatrix} 0.9, & -0.3 \\ 0.4, & -0.7 \end{bmatrix}, \quad F_2 = \begin{bmatrix} 3 \\ 1 \end{bmatrix}, \quad A_2 = \begin{bmatrix} 2.1, & 2.7 \\ 1, & -0.3 \end{bmatrix},$$

$$M_3 = \begin{bmatrix} 0.83, & -0.1 \\ 0.8, & -0.95 \end{bmatrix}, \quad F_3 = \begin{bmatrix} -3 \\ -2 \end{bmatrix}, \quad A_3 = \begin{bmatrix} 1.9, & -0.4 \\ 1, & 0.5 \end{bmatrix}. \tag{6.7}$$

The noise covariance matrices are identical for all components. They are

$$R_\omega = \begin{bmatrix} 0.1, & 0 \\ 0, & 0.1 \end{bmatrix}, \quad R_v = \begin{bmatrix} 0.05, & 0 \\ 0, & 0.03 \end{bmatrix} \tag{6.8}$$

and do not give so much noisy data from which it is possible to expect relatively successful results.

The generated data sample is shown in Fig. 6.18. Switching between components is demonstrated in Fig. 6.19.

Fig. 6.19 Switching the state-space components The mixture switches among three components so that the active component stays the same for a while and only then does it jump to another one

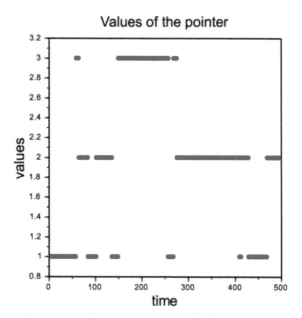

Results The results of the experiment are presented in the following figures. Figure 6.20 demonstrates the correspondence of the estimated state values in relation to the simulated ones. Figure 6.21 shows the estimation of the pointer values, i.e., classification of the data into individual components.

6.4 Case Studies

Experiments with real data provide the final and most important evidence not only about the functionality of the estimating algorithms but they also demonstrate their abilities and restrictions to manage difficult cases with the inherited nonlinearities and nonstationarities.

For this aim, a case study modeling the fuel consumption of a driven car was selected. A dynamic mixture of normal components has been taken but in two variants: one with static components and one with dynamic components. Selected entries of the modeled four-dimensional output y_t include fuel consumption, car speed, gas pedal position, and engine speed.

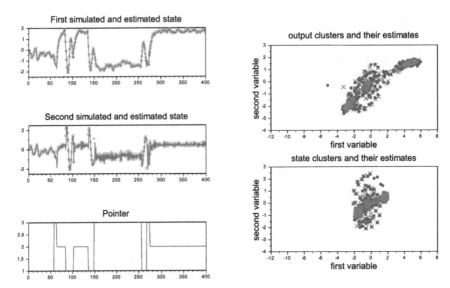

Fig. 6.20 A comparison of simulated and estimated state values. The first two pictures in the *left part* of the figure present the time evolution of simulated values of the state and their estimates. The *left bottom* picture shows the values of the simulated pointer (to explain the behavior of the state variable). The agreement is clearly visible. The *right part* of the figure demonstrates the clusters of the output and the state. Again, it is evident that the clusters are captured very well

6.4.1 Static Normal Components

Initialization Initialization of the component centers should have been performed with the help of a prior data set. An automated form of such a mixture initialization is not available so far, thus expert help was necessary. The expert-based procedure of initialization described in [45] was applied. It consists of the detection of static component centers using the visualization analysis of data. For such a visualization, entries of the output are plotted against each other that give two-dimensional data clusters, as it is shown in Fig. 6.22. Using this procedure, three components have been detected.

Validation procedure Results of the estimation cannot be verified by the comparison of true and estimated parameters because there are no true ones. However, there are some ways how the results can be evaluated.

1. The weighting vector w_t indicates the probabilities of activity of individual components at each time instant t. The point estimate of the pointer, which indicates the active component, is the index of the maximum entry of this vector.

Fig. 6.21 The time evolution of the component weights. The figures show how probabilities of the activity of each of the components evolve in time. It can be seen that they do not express much uncertainty even in the very beginning of the estimation. It can probably be explained by the assumption of known model parameters. It indicates that the estimation of the state is less demanding than the estimation of the whole set of the mixture model parameters

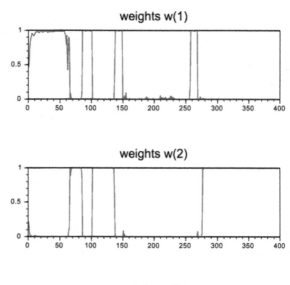

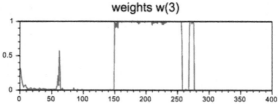

For time $t = 1, 2, \ldots$ we obtain a sequence of labels of active components. If this sequence exhibits a reasonable way of switching the components, there is a good chance that the estimation was successful. On the opposite side, if some component is never active or even if only one component is active, the estimation probably failed.

2. The regression coefficients of the components are coordinates of the centers of the components. If they change at the beginning of the estimation and then they stabilize in some fixed values representing different points (centers) in the data space, we can believe that after the initial search for clusters they found them and the estimation is correct. If some resulting centers are identical or very close to one another, it can indicate that we used too many components and that it is possible to reduce their number.

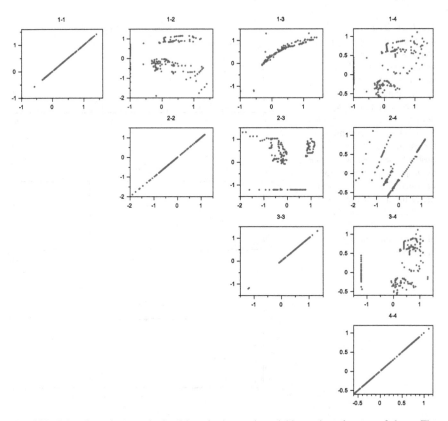

Fig. 6.22 Prior data clusters obtained by *plotting* each variable against the rest of them. The figure has a form of the *upper* triangular matrix, where the *first row* provides clusters of the fuel consumption, first with itself and then with the rest of the variables. The *second row* demonstrates clusters of the car speed and so on. We work with normalized data (i.e., with zero expectations and unit variances), so labels in the graph do not correspond to the real ranges. To obtain prior estimates of component centers, clusters in Figs. 1–2, 2–3, and 3–4 should be considered

3. The most convincing way of verifying the quality of the estimation is to compare the measured and estimated output. The difference between them produces a prediction error. It can be done either graphically—by plotting the output and its estimate or numerically—to evaluate the prediction error on the whole interval of the estimation or on some interval near the end of the estimation. As a suitable numerical characteristic, a sum of squares of the prediction error can serve. Another indicator of successful estimation is the average of the prediction error—in the case of a good estimation it should be zero.

Results The ability of the mixture estimation to model multidimensional real data will be demonstrated in the following figures enriched by an explanation.

The evolution of component weights is demonstrated in Fig. 6.23.

Fig. 6.23 The evolution of component weights. The figures show the time evolution of the corresponding entries of the weighting vector. It can be seen that (i) the components switch in a reasonable way, (ii) the values plotted are mostly near one or zero, which means a resolute decision for the currently active component

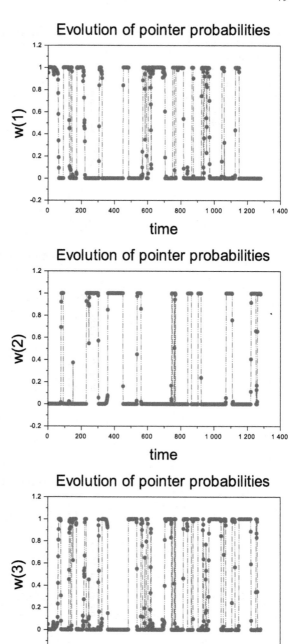

Fig. 6.24 The time
evolution of parameter
estimates of three
components. From these
pictures, a stabilization of
the estimation can be seen.
After an initial search, some
steady state is achieved

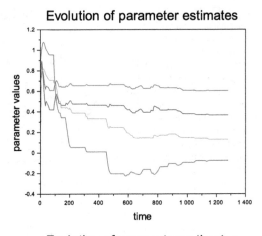

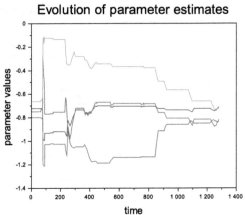

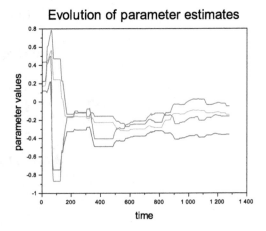

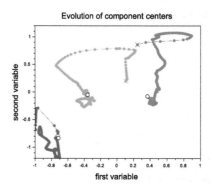
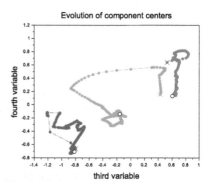

Fig. 6.25 The evolution of estimates in the parameter space. The end of the search is denoted by a *black circle*. The density of points corresponds to the speed of movement

The time evolution of the point estimates of component parameters can be shown in Fig. 6.24.

The evolution of parameter estimates can also be demonstrated in the parameter space, e.g., for variables 1 and 2 (i.e., the fuel consumption and the speed) and separately, 3 and 4 (the gas pedal position and engine speed), see Fig. 6.25. The estimated values of the output variables are compared with the real measured values in Fig. 6.26.

6.4.2 Dynamic Normal Components

To respect the dynamic nature of the data, i.e., the fact that the variables evolve dynamically even in the framework of a single component, we can use dynamic models for components. For the presented experiments, the first-order regression models are taken as components.

Initialization A substantial problem with the dynamic components lays in the initialization of the estimating algorithm. The entries of the constant in the static case correspond to the expectation of components, i.e., with the coordinates of their centers which can be approximately guessed from the initial data sample as it was indicated above. In the dynamic case, we can imitate the static case by starting with parameters as zero matrices and the constant as coordinates of the guessed cluster

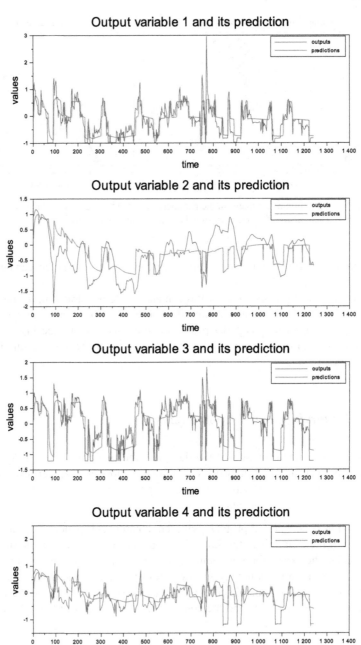

Fig. 6.26 A comparison of the real and estimated values of outputs with static components. The graphs demonstrate a relatively nice coincidence between output variables and their estimates. It is necessary to realize that the demonstrated results are computed using static components. In this way, the changes of the variables can be interpreted only as switching the components. This principle is also apparent from the figures

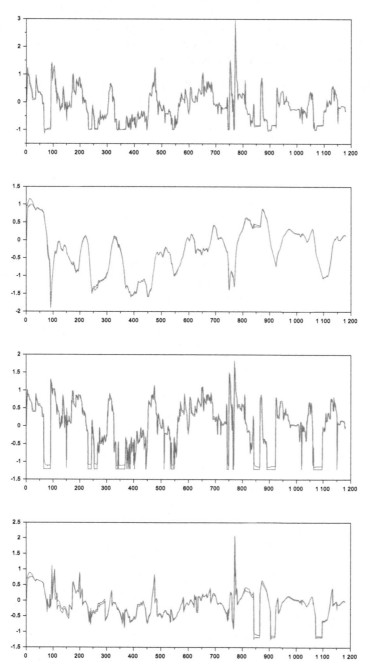

Fig. 6.27 A comparison of the real and estimated values of outputs with dynamic components. A nice coincidence between the output variables and their estimates is observed again. By comparing these results with those from Fig. 6.26, it can be seen that they are more successful

centers. The initial parameters should be with relatively strong belief (introduced through the initial statistics) so that the estimation will not react too quickly if some of the system modes at the very beginning stay hidden.

Results A comparison of real values with output estimates can be seen in Fig. 6.27.

Chapter 7
Appendix A (Supporting Notions)

7.1 Useful Matrix Formulas

Derivative of a determinant For a symmetric matrix A the derivative of its determinant is

$$\frac{\partial}{\partial A}|A| = |A|A^{-1}. \tag{7.1}$$

Derivative of an inversion For a symmetric matrix A the derivative of its inverse is given by the formula

$$\frac{\partial}{\partial A}A^{-1} = -A^{-2}. \tag{7.2}$$

Derivative of a square form For a vector x and a symmetric matrix A it holds

$$\frac{\partial}{\partial A}x'A^{-1}x = -A^{-1}xx'A^{-1}. \tag{7.3}$$

7.2 Matrix Trace

The trace of a square $n \times n$ matrix A is denoted by $\mathrm{tr}\,A$ and is defined as

$$\mathrm{tr}\,A = \sum_{i=1}^{n} A_{i,i}. \tag{7.4}$$

Commutation of matrices under the trace For a trace of a product of matrices it holds

$$\mathrm{tr}\,(ABC) = \mathrm{tr}\,(BCA) = \mathrm{tr}\,(CAB), \tag{7.5}$$

i.e., the matrices commute under the trace.

© The Author(s) 2017
I. Nagy and E. Suzdaleva, *Algorithms and Programs of Dynamic Mixture Estimation*, SpringerBriefs in Statistics, DOI 10.1007/978-3-319-64671-8_7

Derivative of a trace of a square form For a vector x and a symmetric matrix A it holds

$$\frac{\partial}{\partial x} \operatorname{tr}\left(x'Ax\right) = 2Ax. \tag{7.6}$$

The trace for multivariate regression model

The commutation property of the trace is used to convert the multivariate regression model to a form which is suitable for estimation [30]. The exponent of the multivariate normal distribution is modified as follows:

$$-\frac{1}{2}\left(y_t - \theta'\psi_t\right)' r^{-1}\left(y_t - \theta'\psi_t\right) = -\frac{1}{2}\left(\begin{bmatrix} -I \\ \theta \end{bmatrix}' \begin{bmatrix} y_t \\ \psi_t \end{bmatrix}\right)' r^{-1}\left(\begin{bmatrix} -I \\ \theta \end{bmatrix}' \begin{bmatrix} y_t \\ \psi_t \end{bmatrix}\right)$$

$$= -\frac{1}{2}\operatorname{tr}\left(\begin{bmatrix} y_t \\ \psi_t \end{bmatrix}' \begin{bmatrix} -I \\ \theta \end{bmatrix} r^{-1} \begin{bmatrix} -I \\ \theta \end{bmatrix}' \begin{bmatrix} y_t \\ \psi_t \end{bmatrix}\right)$$

$$= -\frac{1}{2}\operatorname{tr}\left(r^{-1} \begin{bmatrix} -I \\ \theta \end{bmatrix}' \begin{bmatrix} y_t \\ \psi_t \end{bmatrix} \begin{bmatrix} y_t \\ \psi_t \end{bmatrix}' \begin{bmatrix} -I \\ \theta \end{bmatrix}\right), \tag{7.7}$$

where in the last equality the commutation is applied. This modification separates data and the parameter part of the distribution, which is a crucial requirement for the recurrent estimation in a closed analytical form.

7.3   Dirac and Kronecker Functions

The **Kronecker function** $\delta\,(i,\,j)$ is defined as follows:

$$\delta\,(i,\,j) = \begin{cases} 1 & \text{for } i = j \\ 0 & \text{otherwise.} \end{cases} \tag{7.8}$$

If a and b are vectors of integers of the same length, the definition of the Kronecker function is analogous

$$\delta\,(a,\,b) = \begin{cases} 1 & \text{for } a_i = b_i, \ \forall i \\ 0 & \text{otherwise.} \end{cases} \tag{7.9}$$

The **Dirac delta function** $\delta\,(t - \tau)$ is defined as a functional acting on continuous functions $g\,(t)$ as follows:

$$\int_{-\infty}^{\infty} g\,(\tau)\,\delta\,(t - \tau)\,d\tau = g\,(t). \tag{7.10}$$

It can also be characterized as a function, which has almost all values equal to zero only in the origin, i.e., for its argument equal to zero it has a value equal to infinity so that it holds $\int_{-\infty}^{\infty} \delta(\tau) d\tau = 1$.

7.4 Gamma and Beta Functions

The **gamma function** is generally defined for complex numbers. Here it is defined for real arguments $x > 0$ as follows:

$$\Gamma(x) = \int_0^{\infty} t^{x-1} \exp\{-t\} dt. \tag{7.11}$$

For the gamma function it holds

$$\Gamma(x+1) = x\Gamma(x). \tag{7.12}$$

The **beta function** can be defined by means of the gamma function in the following way:

$$B(x_1, x_2) = \frac{\Gamma(x_1)\Gamma(x_2)}{\Gamma(x_1 + x_2)} \tag{7.13}$$

and

$$B(x_1 + 1, x_2) = \frac{x_1}{x_1 + x_2} B(x_1, x_2). \tag{7.14}$$

A **multivariate beta function** is a generalization of the beta function for the vector argument $x = [x_1, x_1, \cdots, x_n]$, where $n > 2$

$$B(x) = \frac{\prod_{i=1}^{n} \Gamma(x_i)}{\Gamma\left(\sum_{i=1}^{n} x_i\right)}. \tag{7.15}$$

A **generalized beta function** is defined in relation with the dynamic categorical model. It has matrix arguments

$$x = \begin{bmatrix} x_{1|1} & x_{2|1} & \cdots & x_{n|1} \\ \cdots & & x_{i|j} & \cdots \\ x_{1|n} & & \cdots & x_{n|n} \end{bmatrix}. \tag{7.16}$$

The function is defined as a product of individual multivariate beta functions over all values of the condition (the index after the condition sign |)

$$B\left(x\right) = \prod_{j=1}^{n} B\left(x_{:|j}\right) = \prod_{j=1}^{n} \frac{\prod_{i=1}^{n} \Gamma\left(x_{i|j}\right)}{\Gamma\left(\sum_{i} x_{i|j}\right)}, \tag{7.17}$$

where $x_{:|j}$ means the j-th row of the table.

Proof of the relation (7.15).

The multivariate beta function of $x = [x_1, x_2, \cdots, x_n]$ is defined through the integral

$$B\left(x\right) = \int_{0}^{1} \cdots \int_{0}^{1} \prod_{i=1}^{n} p_i^{x_i-1} \, dp_1 \cdots dp_n, \tag{7.18}$$

where $p_i \geq 0 \; \forall i$ and $\sum_{i=1}^{n} p_i = 1$.

Let us show that it holds

$$B\left(x\right) = \frac{\prod_{i=1}^{n} \Gamma\left(x_i\right)}{\Gamma\left(\sum_{i=1}^{n} x_i\right)}. \tag{7.19}$$

Let us start for $n = 3$. The integrand is

$$p_1^{x_1-1} p_2^{x_2-1} \left(1 - p_1 - p_2\right)^{x_3-1} = p_1^{x_1-1} p_2^{x_2-1} \left[\left(1 - p_1\right)\left(1 - \frac{p_2}{1 - p_1}\right)\right]^{x_3-1} = (*). \tag{7.20}$$

By substituting $\dfrac{p_2}{1 - p_1} = q_2 \rightarrow p_2 = q_2\left(1 - p_1\right)$

$$(*) = p_1^{x_1-1} q_2^{x_2-1} \left(1 - p_1\right)^{x_2-1} \left(1 - p_1\right)^{x_3-1} \left(1 - q_2\right)^{x_3-1} =$$

$$= p_1^{x_1-1} \left(1 - p_1\right)^{x_2+x_3-1} q_2^{x_2} \left(1 - q_2\right)^{x_3-1}. \tag{7.21}$$

After the integration it is obtained

$$B\left(x_1, x_2 + x_3\right) B\left(x_2, x_3\right) =$$

$$= \frac{\Gamma\left(x_1\right) \Gamma\left(x_2 + x_3\right)}{\Gamma\left(x_1 + x_2 + x_3\right)} \frac{\Gamma\left(x_2\right) \Gamma\left(x_3\right)}{\Gamma\left(x_2 + x_3\right)} = \frac{\Gamma\left(x_1\right) \Gamma\left(x_2\right) \Gamma\left(x_3\right)}{\Gamma\left(x_1 + x_2 + x_3\right)}. \tag{7.22}$$

Now, in general

$$\prod_{}^{n-1} p_i^{x_i-1}\left(1-\sum_{}^{n-1} p_i\right)^{x_n-1} = \prod_{}^{n-2} p_i^{x_i-1} p_{n-1}^{x_{n-1}-1}\left(1-\sum_{}^{n-2} p_i - p_{n-1}\right)^{x_n-1} =$$

by substitution $\dfrac{1}{1-\sum^{n-2} p_i} = q_{n-1} \rightarrow p_{n-1} = q_{n-1}\left(1-\sum_{}^{n-2} p_i\right)$

$$= \prod_{}^{n-2} p_i^{x_i-1} q_{n-1}^{x_{n-1}-1}\left(1-\sum_{}^{n-2} p_i\right)^{x_{n-1}-1}\left(1-\sum_{}^{n-2} p_i\right)^{x_n-1}(1-q_{n-1})^{x_n-1} =$$

$$= \prod_{}^{n-2} p_i^{x_i-1}\left(1-\sum_{}^{n-2} p_i\right)^{x_{n-1}+x_n-1} q_{n-1}^{x_{n-1}-1}(1-q_{n-1})^{x_n-1} =$$

$$= \prod_{}^{n-2} p_i^{x_i-1}\left(1-\sum_{}^{n-2} p_i\right)^{x_{n-1}+x_n-1} B\,(x_{n-1},x_n). \qquad (7.23)$$

Recursively it is obtained

$$B\left(x_1,\sum_{2}^{n} x_i\right) B\left(x_2,\sum_{3}^{n} x_i\right) \cdots B\,(x_{n-1},x_n) =$$

$$= \frac{\Gamma\,(x_1)\,\Gamma\left(\sum_2^n x_i\right)}{\Gamma\left(\sum_1^n x_i\right)}\,\frac{\Gamma\,(x_2)\,\Gamma\left(\sum_3^n x_i\right)}{\Gamma\left(\sum_2^n x_i\right)} \cdots \frac{\Gamma\,(x_{n-1})\,\Gamma\,(x_n)}{\Gamma\,(x_{n-1}+x_n)}, \qquad (7.24)$$

which leads to the formula to be proved [31].

7.5 The Bayes Rule

The Bayes rule is used for estimating the parameters of models described by the pdf $f\,(y_t|\psi_t,\Theta)$ by developing the prior/posterior pdf $f\,(\Theta|d\,(t-1)) \rightarrow f\,(\Theta|d\,(t))$. The formula has the form

$$\underbrace{f\,(\Theta|d\,(t))}_{posterior} \propto \underbrace{f\,(y_t|\psi_t,\Theta)}_{model}\,\underbrace{f\,(\Theta|d\,(t-1))}_{prior} \qquad (7.25)$$

with the normalization constant equal to the data predictive pdf $f\,(d_t|d\,(t-1))$.

Remark As the Bayes rule is used recursively in estimation, once for each measured data item, we require the property of **self-reproducibility**. It means that the structure of the prior pdf is not changed when multiplied by the model pdf. Only some of its characteristics (statistics) are allowed to change. This condition guarantees the feasibility of estimation: the complexity of the posterior pdf is not increased during the estimation.

7.6   The Chain Rule

The application of the chain rule, see e.g., [30] can be shown in the following example. For the data collection up to the time $t = 3$, i.e., for $d(3) \equiv \{d_0, d_1, d_2, d_3\}$, where $d_t = \{y_t, u_t\}$ and d_0 is the prior information, it holds

$$f(d(3)|d(0)) = f(d_3|d(2)) f(d_2|d(1)) f(d_1|d(0)). \tag{7.26}$$

7.7   The Natural Conditions of Control

For a controlled model in the form $f(y_t|u_t, d(t-1), \Theta)$ the validity of the following equality is assumed:

$$f(\Theta|u_t, d(t-1)) = f(\Theta|d(t-1)) \tag{7.27}$$

from which, using the Bayes formula, it also follows

$$f(u_t|\Theta, d(t-1)) = f(u_t|d(t-1)). \tag{7.28}$$

These conditions which mean that u_t and Θ are conditionally independent given the past data $d(t-1)$ are called the **natural conditions of control** [30].

7.8   Conjugate Dirichlet Distribution

According to [31], the Dirichlet distribution is conjugate to the categorical model. For the static categorical model, its pdf has the form

$$f(\beta|d(t)) \propto \frac{1}{B(\nu)} \prod_{i=1}^{n} \beta_i^{(\nu_i)_t - 1} \tag{7.29}$$

where $\nu_t = \left[\nu_{1;t}, \nu_{2;t}, \cdots, \nu_{n;t}\right]'$ is a parameter of the distribution, i.e., the statistics for estimation of the parameter β of the model and $B(\nu)$ is the beta function.

For the dynamic categorical model, the Dirichlet pdf is

$$f(\beta|d(t)) = \frac{1}{B(\nu_t)} \prod_{i|j\in(d_t|\psi_t)^*} \beta_{i|j}^{(\nu_{i|j})_t-1}, \qquad (7.30)$$

where the denominator of the normalization constant is the multivariate beta function (7.13). The statistics has the form of a table, where the columns are indexed by the values of the output y_t and the rows by the individual configurations of the regression vector ψ_t. For example, for $\psi_t=\left[y_{t-1}, y_{t-2}\right]$ and two-valued output, the configurations can be ordered as follows: $[1, 1]$, $[1, 2]$, $[2, 1]$, $[2, 2]$. Then the table will be[1]

$\left[y_{t-1}, y_{t-2}\right]$	$y_t = 1$	$y_t = 2$
$[1, 1]$	$\nu_{1|11}$	$\nu_{2|11}$
$[1, 2]$	$\nu_{1|12}$	$\nu_{2|12}$
$[2, 1]$	$\nu_{1|21}$	$\nu_{2|21}$
$[2, 2]$	$\nu_{1|22}$	$\nu_{2|22}$

Individual entries of the table are multi-indexed.[2]

7.8.1 The Normalization Constant of Dirichlet Distribution

The full posterior pdf has the form

$$f(\beta|d(t)) = \frac{1}{B(\nu_t)} \prod_{y|\psi} \beta_{y|\psi}^{(\nu_{y|\psi})_t-1}, \qquad (7.31)$$

where $B(\nu_t)$ is the beta function, generally defined as

$$B(\nu) = \prod_{\psi} \frac{\prod_y \Gamma\left(\nu_{y|\psi}\right)}{\Gamma\left(\sum_y \nu_{y|\psi}\right)}. \qquad (7.32)$$

Details about the beta function can be found in Sect. 7.4.

---

[1] The form of the table is the same as that for parameters β of the model.

[2] Multi-index is an index in the vector form. The notation with | is formal and it is used to stress that some indexes relate to the random variable (before the condition sign), others to the condition (after the condition sign).

7.8.2 Statistics Update with the Conjugate Dirichlet Distribution

Using the Bayes rule (7.25) and according to [31], the prior pdf is chosen in the form of the Dirichlet distribution, which is the conjugate pdf for the considered categorical model

$$f\left(\beta | d\left(t - 1\right)\right) \propto \prod_{y|\psi} \beta_{y|\psi}^{(\nu_{y|\psi})_{t-1}-1}. \tag{7.33}$$

To derive the statistics update [31], the model is written in the product form

$$f\left(y_t | \psi_t, \beta\right) = \prod_{y|\psi} \beta_{y|\psi}^{\delta(y|\psi, \, y_t|\psi_t)}, \tag{7.34}$$

where $\delta\left(y|\psi, \, y_t|\psi_t\right)$ is the multivariate Kronecker function, see Sect. 7.3. It is equal to one if the configuration of $[y|\psi]$ is equal to the configuration of $[y_t|\psi_t]$ (the vectors are equal in all entries) and it is zero otherwise. The product form is formal and only helps to find the recursion for the statistics update.

Substituting the model and the prior Dirichlet pdf into the Bayes rule, it is obtained

$$\prod_{y|\psi} \beta_{y|\psi}^{\delta(y|\psi, \, y_t|\psi_t)} \prod_{y|\psi} \beta_{y|\psi}^{(\nu_{y|\psi})_{t-1}-1} = \prod_{y|\psi} \beta_{y|\psi}^{(\nu_{y|\psi})_{t-1}+\delta(y|\psi, \, y_t|\psi_t)-1} \propto \underbrace{\prod_{y|\psi} \beta_{y|\psi}^{(\nu_{y|\psi})_t-1}}_{\text{posterior}}. \tag{7.35}$$

The above proportionality enables one to easily deduce the statistics update

$$(\nu_{y|\psi})_t = (\nu_{y|\psi})_{t-1} + \delta\left(y|\psi, \, y_t|\psi_t\right) \tag{7.36}$$

for all configurations of $y|\psi$.

7.8.3 The Parameter Point Estimate of the Categorical Model

For the point estimates of parameter β it holds

$$(\hat{\beta}_{i|j})_t = \frac{(\nu_{i|j})_t}{\sum_k (\nu_{k|j})_t}, \quad \forall i | j \in (y_t|\psi_t)^*, \tag{7.37}$$

where entries of the statistics table ν_t are normalized [31].

Proof As the dynamic categorical model can be interpreted as a set of static models, i.e., one for each configuration of the values in the regression vector, here the proof is shown for a static model.

The posterior Dirichlet pdf denoted by $\mathcal{D}i(\nu)$ with the statistics $\nu = [\nu_1, \nu_2, \cdots, \nu_n]$ is

$$f\left(\beta | d\left(t\right)\right) = \mathcal{D}i\left(\nu\right) \propto \prod_{i=1}^{n} \beta_i^{\nu_i - 1}, \quad \text{and} \quad \beta_n = 1 - \sum_{i=1}^{n-1} \beta_i. \tag{7.38}$$

Generally, the point estimate $\hat{\beta}_j$ is taken as the expectation of β_j conditioned by the data comprised in the statistics ν_t. By definition the expectation is equal to

$$E\left[\beta_j | \nu_t\right] = \frac{1}{B\left(\nu_t\right)} \int_0^1 \prod_{i=1}^{n} \beta_i^{\nu_i - 1} d\beta =$$

$$= \frac{1}{B\left(\nu_t\right)} \int_0^1 \beta_j \beta_j^{\nu_j - 1} \prod_{i=1, i \neq j}^{n} \beta_i^{\nu_i - 1} d\beta =$$

$$= \frac{B\left(\left[\nu_1, \nu_2, \cdots, \nu_{j-1}, \nu_j + 1, \nu_{j+1}, \cdots, \nu_n\right]\right)}{B\left(\left[\nu_1, \nu_2, \cdots, \nu_{j-1}, \nu_j, \nu_{j+1}, \cdots, \nu_n\right]\right)} =$$

$$= \frac{\frac{\nu_j}{\sum_{i=1}^{n} \nu_i} B\left(\left[\nu_1, \nu_2, \cdots, \nu_{j-1}, \nu_j, \nu_{j+1}, \cdots, \nu_n\right]\right)}{B\left(\left[\nu_1, \nu_2, \cdots, \nu_{j-1}, \nu_j, \nu_{j+1}, \cdots, \nu_n\right]\right)} = \frac{\nu_j}{\sum_{i=1}^{n} \nu_i} \tag{7.39}$$

and thus

$$\hat{\beta}_{j;t} = \frac{\nu_{j;t}}{\sum_{i=1}^{n} \nu_{i;t}}. \tag{7.40}$$

Here we used the definition of the beta function and the property in an elementary form

$$B\left(\nu_1 + 1, \nu_2\right) = \frac{\nu_1}{\nu_1 + \nu_2} B\left(\nu_1, \nu_2\right). \tag{7.41}$$

7.8.4 Data Prediction with Dirichlet Distribution

For the categorical model with the Dirichlet posterior pdf, the data predictive pdf is

$$f\left(d_t | d\left(t - 1\right)\right) = \int_0^1 f\left(d_t | d\left(t - 1\right), \beta\right) f\left(\beta | d\left(t - 1\right)\right) d\beta =$$

$$= \int_0^1 \beta_{d_t} f\left(\beta | d\left(t - 1\right)\right) d\beta = E\left[\beta_{d_t} | d\left(t - 1\right)\right] = \hat{\beta}_{d_t}$$

which is the d_t-th entry of the expectation of Dirichlet distribution.

7.9   Conjugate Gauss-Inverse-Wishart Distribution

The Gauss-inverse-Wishart (GiW) distribution is conjugate to the normal distribution. Its form (here shown for a scalar case) is

$$f\left(\Theta|d\left(t\right)\right) \propto r^{-0.5\kappa_t} \exp\left\{-\frac{1}{2r}\begin{bmatrix}-1\\\theta\end{bmatrix}'V_t\begin{bmatrix}-1\\\theta\end{bmatrix}\right\}, \tag{7.42}$$

where the square matrix V_t and the number κ_t are statistics for estimation the parameter $\Theta = \{\theta, r\}$ of the normal regression model.

In the case of a static regression model, the form of the prior Gauss-inverse-Wishart pdf is the same as the above one with the regression vector $\psi_t = 1$.

7.9.1   Statistics Update for the Normal Regression Model

The update of the statistics V_t and κ_t of the Gauss-inverse-Wishart pdf during the recursive parameter estimation can be obtained again by substituting the prior and the posterior GiW pdfs (7.42) as well as the normal model pdf into the Bayes rule (7.25), see [8, 30, 31]. After multiplication of the pdfs on the right-hand side shown below for the scalar y_t

$$\frac{1}{\sqrt{r}}\exp\left\{-\frac{1}{2r}\begin{bmatrix}-1\\\theta\end{bmatrix}'\underbrace{\begin{bmatrix}y_t\\\psi_t\end{bmatrix}\begin{bmatrix}y_t\\\psi_t\end{bmatrix}'}_{D_t}\begin{bmatrix}-1\\\theta\end{bmatrix}\right\}\left(\frac{1}{\sqrt{r}}\right)^{\kappa_{t-1}}\exp\left\{-\frac{1}{2r}\begin{bmatrix}-1\\\theta\end{bmatrix}'V_{t-1}\begin{bmatrix}-1\\\theta\end{bmatrix}\right\}$$

$$=\left(\frac{1}{\sqrt{r}}\right)^{\kappa_{t-1}+1}\exp\left\{-\frac{1}{2r}\begin{bmatrix}-1\\\theta\end{bmatrix}'\left(D_t+V_{t-1}\right)\begin{bmatrix}-1\\\theta\end{bmatrix}\right\}$$

$$=\left(\frac{1}{\sqrt{r}}\right)^{\kappa_t}\exp\left\{-\frac{1}{2r}\begin{bmatrix}-1\\\theta\end{bmatrix}'V_t\begin{bmatrix}-1\\\theta\end{bmatrix}\right\}$$

and comparison with the left-hand side of the Bayes rule, one gets the expression for the statistics update (2.8), i.e.,

$$V_t = V_{t-1} + \begin{bmatrix}y_t\\\psi_t\end{bmatrix}\begin{bmatrix}y_t\\\psi_t\end{bmatrix}'$$

$$\kappa_t = \kappa_{t-1} + 1.$$

starting with the initial statistics V_0 and κ_0.

7.9.2 The Parameter Point Estimate of the Regression Model

According to [8, 30, 31] the point estimates of the parameters θ and r of the normal regression model are computed using the partition of the updated matrix statistics V_t in the following way, where the dimensions of sub-matrices are defined by the dimension of the output y_t, see (2.8). It is

$$V_t = \begin{bmatrix} V_y & V'_{y\psi} \\ V_{y\psi} & V_\psi \end{bmatrix},$$ (7.43)

which allows to obtain them as

$$\hat{\theta}_t = V_\psi^{-1} V_{y\psi} \quad \text{and} \quad \hat{r}_t = \frac{V_y - V'_{y\psi} V_\psi^{-1} V_{y\psi}}{\kappa_t}.$$ (7.44)

The proof for the scalar output y_t is provided below [30].

Proof Point estimates are computed as arguments of maximum of the posterior pdf, i.e., for the parameter θ

$$\frac{\partial}{\partial \theta} f(\Theta|d(t)) \rightarrow \frac{\partial}{\partial \theta} r^{-0.5\kappa_t} \exp\left\{ -\frac{1}{2r} \begin{bmatrix} -1 \\ \theta \end{bmatrix}' V_t \begin{bmatrix} -1 \\ \theta \end{bmatrix} \right\} = 0,$$ (7.45)

where $\begin{bmatrix} -1 \\ \theta \end{bmatrix}' V_t \begin{bmatrix} -1 \\ \theta \end{bmatrix} = V_y - 2V_{y\psi}\theta + \theta' V_\psi \theta$. The derivative is

$$r^{-0.5\kappa_t} \exp\left\{ -\frac{1}{2r} \begin{bmatrix} -1 \\ \theta \end{bmatrix}' V_t \begin{bmatrix} -1 \\ \theta \end{bmatrix} \right\} [-2V_{y\psi} + 2V_\psi \theta] = 0$$ (7.46)

and therefore

$$V_\psi \theta = V_{y\psi} \quad \rightarrow \quad \hat{\theta} = V_\psi^{-1} V_{y\psi}.$$ (7.47)

Similarly, for r

$$\frac{\partial}{\partial r} f(\Theta|d(t)) \rightarrow \frac{\partial}{\partial r} r^{-0.5\kappa_t} \exp\left\{ -\frac{1}{2r} \begin{bmatrix} -1 \\ \theta \end{bmatrix}' V_t \begin{bmatrix} -1 \\ \theta \end{bmatrix} \right\} = 0.$$ (7.48)

We differentiate

$$-0.5\kappa_t r^{-.5\kappa_t-1} \exp\left\{-\frac{1}{2r}\begin{bmatrix}-1\\\theta\end{bmatrix}' V_t \begin{bmatrix}-1\\\theta\end{bmatrix}\right\}$$

$$+ r^{-0.5\kappa_t} \exp\left\{-\frac{1}{2r}\begin{bmatrix}-1\\\theta\end{bmatrix}' V_t \begin{bmatrix}-1\\\theta\end{bmatrix}\right\}\frac{1}{2}r^{-2}\left(V_y - 2V_{y\psi}\theta + \theta' V_{\psi}\theta\right) = 0 \quad (7.49)$$

and

$$\kappa_t r^{-1} = r^{-2}\left(V_y - 2V_{y\psi}\theta + \theta' V_{\psi}\theta\right) \quad \rightarrow \quad \kappa_t \hat{r} = V_y - 2V_{y\psi}\theta + \theta' V_{\psi}\theta. \quad (7.50)$$

Substitution the estimate $\hat{\theta} = V_{\psi}^{-1} V_{y\psi}$ instead of θ gives

$$\hat{r} = \frac{V_y - V_{y\psi}' V_{\psi}^{-1} V_{y\psi}}{\kappa_t}. \quad (7.51)$$

7.9.3   The Proximity Evaluation

For the output prediction it holds

$$f\left(d_t|d\left(t-1\right)\right) = \int_{\Theta^*} f\left(y_t|\psi_t, \Theta\right) f\left(\Theta|d\left(t-1\right)\right) d\Theta, \quad (7.52)$$

which can also be expressed as

$$= \int_{\Theta^*} L_t\left(\Theta\right) f\left(\Theta|d\left(0\right)\right) d\Theta, \quad (7.53)$$

where

$$L_t\left(\Theta\right) = \prod_{\tau=1}^{t} f\left(y_\tau|\psi_\tau, \Theta\right)$$

is the likelihood function for the model $f\left(y_t|\psi_t, \Theta\right)$ and $f\left(\Theta|d\left(0\right)\right)$ is the prior pdf. As the likelihood is a generally rather complex function over which it is even necessary to integrate, it is preferable to evaluate the integral in another way, even if approximative. This leads to extreme simplification and better transparency of the solution, while precision remains satisfactory.

This approximate way, e.g., [32] lays in replacing the posterior pdf in (7.52) by a Dirac function, see (7.10), as follows:

$$f\left(\Theta|d\left(t-1\right)\right) \rightarrow \delta\left(\Theta, \hat{\Theta}_{t-1}\right), \tag{7.54}$$

where $\hat{\Theta}_{t-1}$ is the point estimate of the parameter Θ at the time instant $t-1$. It is obtained

$$f\left(d_t|d\left(t-1\right)\right) = f\left(y_t|\psi_t, \hat{\Theta}_{t-1}\right), \tag{7.55}$$

which is equivalent to substituting the point estimate $\hat{\Theta}_{t-1}$ instead of Θ to avoid working with the full posterior pdf for Θ.

Remark Similarly, it is possible to deal with the unknown future output d_{t+1} predicting d_{t+2}

$$f\left(d_{t+2}|d\left(t\right)\right) = \int f\left(d_{t+2}|d_{t+1}, d\left(t\right)\right) f\left(d_{t+1}|d\left(t\right)\right) dd_{t+1} \doteq f\left(d_{t+2}|\hat{d}_{t+1}, d\left(t\right)\right), \tag{7.56}$$

where $\hat{d}_{t+1}$ is the point estimate of d_{t+1}.

Chapter 8
Appendix B (Supporting Programs)

This chapter represents Scilab codes of supporting programs necessary for using the algorithms from Chap. 5. Part of them are the codes of simulating data using a corresponding type of components of the dynamic mixture. The rest are the auxiliary programs simplifying the performance of the main estimation program and the subroutines (see Chap. 5).

8.1 Simulation Programs

For simulating a data sample using a dynamic mixture, first it is necessary to generate the values of the pointer variable, which expresses switching the mixture components via the transition table. Having this table and the initial pointer value, the pointer values are generated. According to its values at the actual time instants the data sample is simulated from the corresponding component.

8.1.1 The Simulation of Pointer Values

Let us consider a discrete random variable c_t described by the dynamic categorical model (3.2) with the parameter α, which is a square matrix of the dimension equal to the number of different values of the pointer c_t. Each entry of α is denoted by $\alpha_{i|j}$, where $i|j$ is the multi-index and $\alpha_{i|j}$ is the probability of a transition from the $c_{t-1} = j$ to the $c_t = i$. For instance, the table can be presented as

© The Author(s) 2017

I. Nagy and E. Suzdaleva, *Algorithms and Programs of Dynamic Mixture Estimation*, SpringerBriefs in Statistics, DOI 10.1007/978-3-319-64671-8_8

	$c_t = 1$	$c_t = 2$	$\cdots$
$c_{t-1} = 1$	$\alpha_{1|1}$	$\alpha_{2|1}$	$\cdots$
$c_{t-1} = 2$	$\alpha_{1|2}$	$\cdots$	$\cdots$
$\cdots$			

where all rows are normalized to a sum that is equal to one. In this table c_t represents the row index and c_{t-1} the column one. Thus, the matrix defined by the table is transposed (in matrices the first index is row and the second is the column one).

The previous pointer value c_{t-1} is necessary to start the pointer generation according to the following formula:

$$c_t = \text{sum}\,(rand > \text{cumsum}\,(row)) + 1, \tag{8.1}$$

where

- row is the chosen row of the table with the sum of the entries equal to one;
- cumsum(row) creates borders of the division of a unit interval into subintervals with lengths equal to probabilities from the row;
- The command > assigns one to the subintervals whose upper border is lower than the generated uniform random number $rand$.

The rest is zero. Summing these values gives the order of the interval to which the value of $rand$ belongs. This value has the probability equal to the length of the corresponding subinterval which is the probability of the corresponding entry of the row.

Then the data sample is simulated using the current c_t and the corresponding components, see the programs below. The descriptions of the commands are inserted directly in the programs.

Remark Surely this scheme can be used not only for the pointer value generation but also for the general simulation of a discrete data sample from a categorical model.

8.1.2   The Simulation of Mixture with Regression Components

```
//// Simulation of two dimensional static regression mixture
////                     with dymamic pointer
// -----------------------------------------------------------------------
// basic simulation examples
chdir(get_absolute_file_path("E1Sim.sce")),mode(0),funcprot(0);
exec("Intro.sce",-1)

nd=1500;                                    // number of data
```

```
nc=3;                                    // number of components
ny=2;                                    // dimension of the output

pp=[.69 .03 .07 .07 .14                  // pointer parameters
    .01 .58 .06 .29 .06
    .05 .02 .62 .19 .12
    .06 .02 .12 .59 .06
    .11 .17 .11 .06 .55];
pp=fnorm(pp(1:nc,1:nc),2);
cp=cumsum(pp,2);
// here you can define your own component parameters:
ths=list();
ths(1)=[2 -2]';
ths(2)=[1 5]';
ths(3)=[8 1]';
ths(4)=[5 4]';
ths(5)=[-1 2]';
amp=.3;
cvs=list();
cvs(1)=amp*[1 .2;.2 2];
cvs(2)=amp*[2 -.9;-.9 1];
cvs(3)=amp*[1 -.3;-.3 1];
cvs(4)=amp*[2 .2;.2 3];
cvs(5)=amp*[3 .2;.2 3];
// -------------------------------------
ct(1)=1;                                 // initial component
// SIMULATION
for t=2:nd
  ct(t)=sum(rand(1,1,'unif')>cp(ct(t-1),:))+1;       // component generation
  yt(:,t)=ths(ct(t))+cvs(ct(t))*rand(ny,1,'norm');   // data generation
end

// generation of prior classified data
yi=[]; ni=5;                             // no. of data from each comp.
for i=1:nc
  j=find(ct==i);                         // data from component i
  ji=j(10+(1:ni));                       // selected from data sample
  yi=[yi [yt(:,ji); i*ones(1,ni)]];      // stored in yi
end

// RESULTS
set(figure(1),'position',[140 50 500 500])
set(gcf(),'background',8);
plot(yt(1,:),yt(2,:),'.')
title('Data')

set(figure(2),'position',[660 50 500 500])
set(gcf(),'background',8);
plot(ct,'.')
title('Clusters')
set(gca(),'data_bounds',[1 nd .9 nc+.1])

// SAVING THE DATA SAMPLE (the name is compulsory)
als=pp; ncs=nc;
save('_data/datRegEx','yt','ct','yi','ths','cvs','als','ncs')
```

8.1.3   The Simulation of Mixture with Discrete Components

```
// Simulation of two dimensional static discrete mixture
//                  with dymamic pointer
//   - structure
//   pointer     f(c|c1,al)   al: matrix 3x3
//   components  f(y1,y2|c,par_c)  par: table 3x4  with head
//                                    y1,y2=1,1  1,2  2,1  2,2
//                                    c=1  |    |    |    |
//                                    ----------------
//                                    c=2  |    |    |    |
//                                    ----------------
//                                    c=3  |    |    |    |
// -----------------------------------------------------------------------
exec("Intro.sce",-1), mode(0)

nd=1000;                  // number of data
ky=[2 2];                 // number of values of y(1,:) and y(2,:)
nc=3;                     // number of components
     // for nc > 3, the parameters al and p must be extended

// transition table of the pointer
al=[7 1 2 2 1
    1 8 2 2 1
    1 2 8 2 1
    2 2 1 5 1
    1 2 2 1 8];
al=fnorm(al(1:nc,1:nc),2); // resetting number of components (at 2)

// parameters of components
p=list();
par=[130 3 4 2]; p(1)=par/sum(par);     // the length of parameters
par=[2 120 1 3]; p(2)=par/sum(par);     //  must be equal to
par=[1 1 3 140]; p(3)=par/sum(par);     //  prod(ky) !!!!
par=[1 1 100 1]; p(4)=par/sum(par);
par=[1 200 3 8]; p(5)=par/sum(par);


c1=2;
ct=ones(1,nd);
yt=ones(2,nd);

// SIMULATION
for t=2:nd
  pr=al(c1,:);
  c=sum(rand(1,1,'unif')>cumsum(pr))+1; // generation of active component
  yy=sum(rand(1,1,'unif')>cumsum(p(c)))+1; // generation of data from act.comp.
  yt(:,t)=row2psi(yy,ky)';              // transformation to y(1) and y(2)
  c1=c;                                 // remember as past component
  ct(t)=c;                              // remember for graph
end

// generation of prior classified data
yi=[]; ni=5;                            // no. of data from each comp.
```

```
for i=1:nc
  j=find(ct==i);                    // data from component i
  ji=j(10+(1:ni));                  // selected from data sample
  yi=[yi [yt(:,ji); i*ones(1,ni)]]; // stored in yi
end

// RESULTS
scf(1);
set(gcf(),'background',8);
plot(yt(1,:)+.01*rand(1,nd,'norm'),yt(2,:)+.01*rand(1,nd,'norm'),'.:')
title('Data clusters','fontsize',4)

scf(2);
plot(1:nd,ct,'.')
title('Component switching','fontsize',4)
set(gca(),'data_bounds',[1 nd .9 nc+.1])

// SAVING THE DATA SAMPLE (the name is compulsory)
als=al; ps=p; ncs=nc; kys=ky;
save('_data/datDisEx','yt','ct','yi','als','ps','ncs','kys')
```

8.1.4 The Simulation of Mixture with State-Space Components

```
//// Simulation of two dimensional state-space mixture
////                    with dynamic pointer
// --------------------------------------------------------------------------
exec("Intro.sce",-1)

nd=1000;                            // number of data
nc=5;                               // number of components <= 5
cF=.5;                              // proximity of initial components

M=list(); A=list(); F=list(); Rw=list(); Rv=list();

// component model parameters
M(1)=[.3 -.3; .5 .1]; A(1)=[.9 .4;1 -.4]; F(1)=cF*[-1 2]';
Rw(1)=[1 0; 0 1]*.1; Rv(1)=[.1 0;0 .1]*.1;

M(2)=[.05 -.3; .4 -.1]; A(2)=[2.1 2.8;1 -.3]; F(2)=cF*[3 1]';
Rw(2)=[1 0; 0 1]*.1; Rv(2)=[.2 0;0 .2]*.1;

M(3)=[.6 -.1; .8 -.5]; A(3)=[1.9 -.4;1 .5]; F(3)=cF*[-3 -2]';
Rw(3)=[1 0; 0 1]*.1; Rv(3)=[.5 0;0 .3]*.1;

M(4)=[.2 -.5; .2 -.9]; A(4)=[.9 -.8;1 1.5]; F(4)=cF*[3 -2]';
Rw(4)=[.1 0; 0 .1]*.1; Rv(4)=[.05 0;0 .8]*.1;

M(5)=[.9 -.01; .08 -.3]; A(5)=[2.9 -.04;.1 1.5]; F(5)=cF*[-1 3]';
Rw(5)=[.1 0; 0 1]*.1; Rv(5)=[1.5 0;0 .6]*.1;

// pointer model parameters
als=[10 1 1 1 1
      1 10 1 1 1
      1 1 10 1 1
```

```
      1 1 1 10 1
      1 1 1 1 10];
als=fnorm(als(1:nc,1:nc),2);

x=zeros(2,nd);                               // declarations
yt=zeros(2,nd);
ct=zeros(1,nd);

ct(1)=1;                                     // initial past component
// SIMULACE
for t=2:nd
  c=sum(rand(1,1,'unif')>cumsum(als(ct(t-1),:)))+1; // generation of act. comp.
  x(:,t)=M(c)*x(:,t-1)+F(c)+Rw(c)*rand(2,1,'norm'); // generation of state
  yt(:,t)=A(c)*x(:,t)+Rv(c)*rand(2,1,'norm');       // generation of output
  ct(t)=c;                                   // stor for graph
end

// RESULTS
cla
set(scf(1),'position',[600 50 500 500])
title('Output and state variables','fontsize',4)
subplot(211)
plot(yt')
xlabel('time','fontsize',4)
ylabel('variables','fontsize',4)
set(gca(),'margins',[0.2,0.2,0.2,0.2])
subplot(212)
plot(x')
xlabel('time','fontsize',4)
ylabel('variables','fontsize',4)
set(gca(),'margins',[0.2,0.2,0.2,0.2])

set(scf(2),'position',[700 100 500 500])
title('Output and state clusters','fontsize',4)
subplot(211)
plot(yt(1,:),yt(2,:),'.')
xlabel('first variable','fontsize',4)
ylabel('second variable','fontsize',4)
set(gca(),'margins',[0.2,0.2,0.2,0.2])
subplot(212)
plot(x(1,:),x(2,:),'.')
xlabel('first variable','fontsize',4)
ylabel('second variable','fontsize',4)
set(gca(),'margins',[0.2,0.2,0.2,0.2])

set(scf(3),'position',[100 50 500 500])
title('Values of the pointer','fontsize',4)
plot(ct,'.')
xlabel('time','fontsize',4)
ylabel('pointer values','fontsize',4)
set(gca(),'margins',[0.2,0.2,0.2,0.2])
set(gca(),'data_bounds',[1 nd .9 nc+.1])

// SAVING THE DATA SAMPLE (the name is compulsory)
ncs=nc;
save('_data/datStaEx','yt','ct','als','ncs','M','A','F','Rw','Rv')
```

8.2 Supporting Subroutines

8.2.1 Scilab Start Settings

The subsequent procedure *Intro* contains the initial settings of a start of Scilab. This procedure is used in the beginning of each of the presented procedures. It is only an auxiliary option and is not necessary in the presented form. However, it is rather helpful for a Scilab 5.5.2 user. Comments are provided inside the program.

Program

```
// General introduction to SciLab run
// ====================================
mode(0)                       // behaves like an exec-file
clear, clearglobal()          // clear variables
xdel(winsid())                // clear graphs
funcprot(0)                   // renaming functions without echo
warning('off')                // suppress warnings

[u,t,n]=file();               // find working directory
chdir(dirname(n(1)));         // set working directory

getd '_func/';                // functions from common library

clc                           // clear console
```

8.2.2 The Point Estimation of a Normal Regression Model

The following program provides the computation of the point estimates of the parameter θ and r of the normal regression model using the updated statistics V_t according to (2.8) and (2.10) from Sect. 2.1.

Program

```
function [th,s2]=v2thN(v,m)
  // [th,s2]=v2thn(v,m)   computation of par. point estimates
  //                      from normalized information matrix
  // v    information matrix (divided by number of samples)
  // m    dimension of y
  // th   regression coefficients
  // s2   noise cobariance estimate

  if argn(2)<2   // check for number of input arguments
    m=1;
  end

  // partitioning of information matrix
```

```
vy=v(1:m,1:m);
vyf=v(m+1:$,1:m);
vf=v(m+1:$,m+1:$);

// computation of point estimates
th=inv(vf+1e-12*eye(vf))*vyf;
s2=(vy-vyf'*th);
endfunction
```

8.2.3   The Value of a Normal Multivariate Distribution

The proximity of a data item from the component can be expressed as a value of
the normal distribution with the substituted data item and the point estimate of the
parameters. The value of the normal multivariate distribution

$$N_d\,(m,r) = \frac{1}{(2\pi)^{-0.5n}\,|r|^{-0.5}}\,\exp\left\{-0.5\,(d-m)'\,r^{-1}\,(d-m)\right\} \qquad (8.2)$$

is computed in the program given below.

Program

```
function [p,Lp]=GaussN(x,m,R)
   // [p Lp]=GaussN(x,m,R)    value of multivariate Gaussian pdf

   // p          probability
   // Lp         logarithm of prob.
   // x          realization
   // m          expectation
   // R          covariance matrix

   x=x(:);            // column vector
   m=m(:);            // column vector

   n=max(size(R));
   Lp=-.5*(n*log(2*%pi)+log(det(R)));
   ex=(x-m)'*inv(R+1e-8*eye(n,n))*(x-m);
   Lp=Lp-.5*ex;
   p=exp(Lp);
endfunction
```

8.2.4   Discrete Regression Vector Coding

When dealing with discrete models with regression vectors composed of several
variables, it is advantageous to introduce a new coding for the regression vector. A
specific configuration of the regression vector is assigned a number corresponding
to the decimal evaluation of the regression vector in the basis given by the numbers
of different values of the variables in the regression vector (increased by one), e.g.,

for three variables $\psi = [d_1, d_2, d_3]$ and $d_i \in \{1, 2\}$ the configuration $\psi = [2, 1, 2]$ is assigned a value of 6. This coding is closely bound to the discrete model table

ψ_t	$y_t = 1$	$y_t = 2$	N^o of reg.vec
[1, 1, 1]			1
[1, 1, 2]			2
[1, 2, 1]			3
[1, 2, 2]			4
[2, 1, 1]			5
[2, 1, 2]			6
...			...

where coding the regression vector is given by the order number of the row of the model table.

The following two programs perform converting from the regression vector to the corresponding number of the table row and backward.

Program

```
function i=psi2row(x,b)
  // i=psi2row(x,b)   i is row number of a model table with
  //                  the regression vector x with the base b;
  //                  elements of x(i) are 1,2,...,nb(i)
  // it is based on the relation
  //     i=b(n-1)b(n-2)...b(1)(x(n)-1)+...+b(1)(x(2)-1)+x(1)

  n=length(x);
  if argn(2)<2, b=2*ones(1,n); end
  b=[b(2:n) 1];
  i=0;
  for j=1:n
    i=(i+x(j)-1)*b(j);
  end
  i=i+1;
endfunction

function x=row2psi(i,b)
  // x=row2psi(i,b)   generates a discrete regression vector
  //                  with a base b that corresponds to the
  //                  i-th row of a table of the discrete model
  // it is based on the relation
  //     i=b(n-1)b(n-2)...b(1)(x(n)-1)+...+b(1)(x(2)-1)+x(1)

  if isscalar(b),
    n=fix(log(i+1)/(log(b)))+1;
    b=b*ones(1,n);
  else
    n=length(b);
  end
  if i>prod(b)
    disp('ERROR: The row number is too big')
    return
  end
  i=i-1;
  for j=1:n
```

```
    i=i/b(n-j+1);
    x(n-j+1)=round(((i-fix(i))*b(n-j+1)+1);
    i=fix(i);
  end
  x=x(:)';
endfunction
```

8.2.5   Kalman Filter

The Kalman filter implemented according to Sect. 2.3.1 with its detailed description
is listed in the following program.

Program

```
function KF=KFilt(xt,yt,A,C,F,Rw,Rv,Rx,ie)
  // [xt,ey,Rx,yp]=KFilt(xt,yt,A,C,F,Rw,Rv,Rx,ie)
  // Kalman filter for the state-space model in the following form
  //                    xt = A*xt + F
  //                    yp = C*xt
  // ... first compute prediction and then filtration, i.e.,
  //              t-1|t-1  -->  t|t-1  -->  t|t
  // KF.xt    state estimate
  // KF.yp    predicted output
  // KF.Ry    output covariance
  // KF.ey    prediction error
  // KF.Rx    state covariance matrix
  // yt    data sample
  // A,C,F model parameters
  // Rw    state covariance
  // Rv    output covariance
  // i.e.,    indicator ie=0 => only state prediction is computed
  //
  if argn(2)<9, ie=1; end

  // Prediction
  xt=A*xt+F;                          // time updt of state
  Rx=Rw+A*Rx*A';                      // time updt of state covariance

  // Filtration
  yp=C*xt;                            // data prediction
  Ry=Rv+C*Rx*C';                      // output prediction covariance
  Ryy=Ry+1e-8*eye(Ry);
  Rx=Rx-Rx*C'*inv(Ryy)*C*Rx;           // data updt of  state covariance
  ey=yt-yp;                           // prediction error
  if ie==0                            // skip filtration
    ey=zeros(ey);
  else
    KG=Rx*C'*inv(Rv);                  // Kalman gain
    xt=xt+KG*ey;                       // data updt of state
  end

  KF.xt=xt;
  KF.yp=yp;
```

```
  KF.Ry=Ry;
  KF.ey=ey;
  KF.Rx=Rx;
endfunction
```

8.2.6 Matrix Upper–Lower Factorization

In a scalar case, a random variable has the variance r and the standard deviation $s = \sqrt{r}$. For a random vector, we have the covariance matrix r and the role of the standard deviation s plays the so-called UL-factor U of the matrix, which is an upper triangular matrix, for which it holds

$$r = U.U'. \tag{8.3}$$

The factorization is performed by the following program.

Program

```
function u=uut(m)
// u=uut(m)    m=uu' UL-factorization of sym.poz.def. matrix
// u   upper triangular matrix
// m   symmetric poz.def. square matrix

n=max(size(m));
for i=n:-1:1
  s=0;
  for j=(i+1):n
    s=s+u(i,j)^2;
  end
  u(i,i)=sqrt(m(i,i)-s);
  for k=(i-1):-1:1
    s=0;
    for j=(i+1):n
      s=s+u(i,j)*u(k,j);
    end
    u(k,i)=(m(k,i)-s)/u(i,i);
  end
end
endfunction
```

8.2.7 Transition Table Normalization

The auxiliary program presented below provides the normalization of the transition table of a categorical model so that the sum of all probabilities in each row will be equal to 1.

Program

```
function fn=fnorm(f,i)
  // fn=fnorm(f,i)    normalization of probabilistic table
  // fn   normalized table
  // f    table
  // i    direction i=1 norm columns, i=2 norm rows

  if argn(2)==1,
    sf=sum(f);
    fn=f/sf;
  else
    [m n]=size(f);
    if i==1
      f1=sum(f,1);
      fn=f./(ones(m,1)*f1);
    else
      f2=sum(f,2);
      fn=f./(f2*ones(1,n));
    end
  end
endfunction
```

8.2.8   The Approximation of Normal Pdfs by a Single Pdf

The approximation of several normal pdfs by a single normal pdf based on minimizing the Kerridge inaccuracy [41] (see 4.19 and 4.20) is fulfilled with the help of the following program.

Program

```
function [x,R]=connect(w,Est)
  // [x,R]=connect(w,Est)    connection Gauss pdfs into one
  //                         Gauss = prod(Gauss_i^wi)
  // Est(i).x   state estimate for i-th component
  // Est(i).cv  covariance matrix for i-th component

  nc=length(w);
  sr=0;
  for i=1:nc
    sr=sr+w(i)*inv(Est(i).Rx);
  end
  R=inv(sr);
  sx=0;
  for i=1:nc
    sx=sx+w(i)*inv(Est(i).Rx)*Est(i).xt;
  end
  x=R*sx;
endfunction
```

References

1. S. Haykin, *Neural Networks: A Comprehensive Foundation* (Macmillan, New York, 1994)
2. B.-J. Park, Y. Zhang, D. Lord, Bayesian mixture modeling approach to account for heterogeneity in speed data. Transp. Res. Part B Methodol. **44**(5), 662–673 (2010)
3. Y. Zou, Y. Zhang, D. Lord, Analyzing different functional forms of the varying weight parameter for finite mixture of negative binomial regression models. Anal. Methods Accid. Res. **1**, 39–52 (2014)
4. Y. Xiong, F.L. Mannering, The heterogeneous effects of guardian supervision on adolescent driver-injury severities: a finite-mixture random-parameters approach. Transp. Res. Part B Methodol. **49**, 39–54 (2013)
5. J. Yu, A nonlinear kernel Gaussian mixture model based inferential monitoring approach for fault detection and diagnosis of chemical processes. Chem. Eng. Sci. **68**(1), 506–519 (2012)
6. J. Yu, A particle filter driven dynamic Gaussian mixture model approach for complex process monitoring and fault diagnosis. J. Process Control **22**(4), 778–788 (2012)
7. J. Yu, Fault detection using principal components-based Gaussian mixture model for semiconductor manufacturing processes. IEEE Trans. Semicond. Manuf. **24**(3), 432–444 (2011)
8. M. Kárný, J. Kadlec, E.L. Sutanto. Quasi-Bayes estimation applied to normal mixture, in *Preprints of the 3rd European IEEE Workshop on Computer-Intensive Methods in Control and Data Processing, CMP'98 /3./*, ed. by J. Rojíček, M. Valečková, M. Kárný, K. Warwick (Prague, CZ, 07.09.1998–09.09.1998), pp. 77–82
9. B.B. Alhaji, H. Dai, Y. Hayashi, V. Vinciotti, A. Harrison, B. Lausen, Bayesian analysis for mixtures of discrete distributions with a non-parametric component. J. Appl. Stat. **43**(8), 1369–1385 (2016)
10. M. Kárný, Recursive estimation of high-order Markov chains: approximation by finite mixtures. Inf. Sci. **326**(1), 188–201 (2016)
11. G. McLachlan, D. Peel, *Finite Mixture Models*, 1st edn. (Wiley-Interscience, 2000)
12. S. Frühwirth-Schnatter, *Finite Mixture and Markov Switching Models*, 2nd edn. (Springer, New York, 2006)
13. O. Cappé, E. Moulines, T. Ryden, *Inference in Hidden Markov Models*, Springer Series in Statistic (Springer-Verlag, 2005)
14. K. Mengersen, C. Robert, M. Titterington, *Mixtures: Estimation and Applications*, 1st edn. (Wiley, 2011)
15. B.H. Iswanto, *New Algorithms for Learning of Mixture Models and Their Application for Classification and Density Estimation* (Logos Verlag, 2005)
16. M.R. Gupta, Y. Chen, *Theory and Use of the EM Method. (Foundations and Trends(r) in Signal Processing)* (Now Publishers Inc, 2011)

© The Author(s) 2017

I. Nagy and E. Suzdaleva, *Algorithms and Programs of Dynamic Mixture Estimation*, SpringerBriefs in Statistics, DOI 10.1007/978-3-319-64671-8

17. G. McLachlan, T. Krishnan, *The EM Algorithm and Extensions*, 2nd edn. (Wiley-Interscience, 2008)
18. O. Boldea, J.R. Magnus, Maximum likelihood estimation of the multivariate normal mixture model. J. Am. Stat. Assoc. **104**(488), 1539–1549 (2009)
19. J.A. Cuesta-Albertos, C. Matran, A. Mayo-Iscar. Robust estimation in the normal mixture model based on robust clustering. J. R. Stat. Soc. Ser. B Stat. Methodol. **70**(Part 4), 779–802 (2008)
20. H. Wang, B. Luo, Q. Zhang, S. Wei, Estimation for the number of components in a mixture model using stepwise split-and-merge EM algorithm. Pattern Recognit. Lett. **25**(16), 1799–1809 (2004)
21. H. Zeng, Y. Cheung, Learning a mixture model for clustering with the completed likelihood minimum message length criterion. Pattern Recognit. **47**(5), 2011–2030 (2014)
22. S.K. Ng, G.J. McLachlan, Mixture models for clustering multilevel growth trajectories. Comput. Stat. Data Anal. **71**, 43–51 (2014)
23. P. Schlattmann, *Medical Applications of Finite Mixture Models (Statistics for Biology and Health)* (Springer, 2009)
24. C.A. McGrory, D.M. Titterington, Variational Bayesian analysis for hidden Markov models. Aust. N. Z. J. Stat. **51**, 227–244 (2009)
25. V. Šmídl, A. Quinn, *The Variational Bayes Method in Signal Processing* (Springer, 2005)
26. Z. Ghahramani, G.E. Hinton, Variational learning for switching state-space models. Neural Comput. **12**(4), 831–864 (2000)
27. S. Frühwirth-Schnatter, Fully bayesian analysis of switching gaussian state space models. Ann. Inst. Stat. Math. **53**(1), 31–49 (2001)
28. R. Chen, J.S. Liu, Mixture kalman filters. J. R. Stat. Soc. B **62**, 493–508 (2000)
29. A. Doucet, C. Andrieu, Iterative algorithms for state estimation of jump Markov linear systems. IEEE Trans. Signal Process. **49**(6), 1216–1227 (2001)
30. V. Peterka, Bayesian system identification, in *Trends and Progress in System Identification*, ed. by P. Eykhoff (Pergamon Press, Oxford, 1981), pp. 239–304
31. M. Kárný, J. Böhm, T.V. Guy, L. Jirsa, I. Nagy, P. Nedoma, L. Tesař, *Optimized Bayesian Dynamic Advising: Theory and Algorithms* (Springer, London, 2006)
32. I. Nagy, E. Suzdaleva, M. Kárný, T. Mlynářová, Bayesian estimation of dynamic finite mixtures. Int. J. Adapt. Control Signal Process. **25**(9), 765–787 (2011)
33. I. Nagy, E. Suzdaleva, Mixture estimation with state-space components and Markov model of switching. Appl. Math. Model. **37**(24), 9970–9984 (2013)
34. J. Berger, *Statistical Decision Theory and Bayesian Analysis* (Springer, New York, 1985)
35. A. Zellner, *An Introduction to Bayesian Inference in Econometrics*, 1st edn. (Wiley-Interscience, 1996)
36. M. West, J. Harrison, *Bayesian Forecasting and Dynamic Models*, 2nd edn. (Springer, 1997)
37. R.E. Kalman, A new aproach to linear filter and prediction theory. J. Basic Eng. D **82**, 35–45 (1960)
38. H.W. Sorenson (ed.), *Kalman Filtering: Theory and Application* (IEEE, 1960)
39. M. Grewal, A. Andrews, *Kalman Filtering: Theory and Practice Using MATLAB*, 2nd edn. (Wiley, 2001)
40. D.M. Titterington, A.F.M. Smith, U.E. Makov, *Statistical Analysis of Finite Mixture Distributions* (Applied probability and statistics (Wiley, Wiley series in probability and mathematical statistics, 1985)
41. D. Kerridge, Inaccuracy and inference. J. R. Stat. Soc. B **23**, 284–294 (1961)
42. H. Steinhaus, Sur la division des corp materiels en parties. Bull. Acad. Polon. Sci. **IV**(C1.III), 801–804 (1956)
43. S. Lloyd, Least squares quantization. PCM. IEEE Trans. Inf. Theory **28**, 129–137, (1982) (Originally as an unpublished Bell laboratories Technical Note, 1957)
44. A.K. Jain, Data clustering: 50 years beyond K-means. Pattern Recognit. Lett. **31**(8), 651–666 (2010)

45. E. Suzdaleva, I. Nagy, T. Mlynářová, Expert-based initialization of recursive mixture estimation, in *Proceedings of 2016 IEEE 8th International Conference on Intelligent Systems IS'2016*, pp. 308–315, (Sofia, Bulgaria, 2016)
46. E. Frank, M. A. Hall, I. H. Witten, The WEKA Workbench, in Online Appendix for *"Data Mining: Practical Machine Learning Tools and Techniques"*, 4th edn. (Morgan Kaufmann, 2016)

CPSIA information can be obtained
at www.ICGtesting.com
Printed in the USA
LVOW02s0737270817
546500LV00002B/2/P